LIVING **ETHICS** SERIES

Laudato Si': A Critique

Pope Francis' Encyclical Letter On the Care of Our Common Home

Dr John I Fleming
Foreword by Professor John Ozolins

Modotti Press
AN IMPRINT OF CONNOR COURT PUBLISHING

Published in 2016 as a part of the Living Ethics Series

Published by Modotti Press, an imprint of Connor Court Publishing Pty Ltd

Copyright © Fr John Fleming

ALL RIGHTS RESERVED. This book contains material protected under International and Federal Copyright Laws and Treaties. Any unauthorised reprint or use of this material is prohibited. No part of this book may be reproduced or transmitted in any form or by any means, electronic or mechanical, including photocopying, recording, or by any information storage and retrieval system without express written permission from the publisher.

PO Box 224W
Ballarat VIC 3350
sales@connorcourt.com
www.connorcourt.com

ISBN: 9781925138931 (pbk)

Cover design by Ian James

Printed in Australia

Bible quotations taken from the Revised Standard Version, used with permission.

Editorial note: In this book traditional inclusive language is used such that the word "man" includes both the male and the female as the context indicates.

Dedicated to
Lucy, Bijou, Winky, and Walter

Previous Volume in the
LIVING ETHICS SERIES

Dignitas Personae Explained
The Catholic Church's teaching on reproductive and related technologies

Dr John Fleming
Foreword by Dr John Haas

Available from:
www.connorcourt.com

CONTENTS

Foreword — 7
Preface — 11

CHAPTER 1: Interpreting the Encyclical — 13
But what is an "Encyclical Letter"? — 13
How are we to interpret a particular encyclical? — 16

CHAPTER 2: The ecological crisis of the twentieth century — 21
The Lynn White Intervention — 22
The Enlightenment, Secularists, and the Environment — 25
Magister mundi sum! – I am the master of the universe! — 26
The impact of secular political philosophy on theology — 33
Secularist approaches to environmental sustainability — 45
The meaning of gender equality — 47
Sustainable Development — 48
Engagement by the Catholic Church with the secular world — 49
Further philosophical considerations — 51

CHAPTER 3: What is Catholic social teaching? — 58
What then is Catholic social teaching? — 63
Another formulation of the basis of Catholic social teaching — 67
The limits of Catholic social teaching — 69
Principles and Values — 70
The Church's right to speak — 71
Conclusion — 80

CHAPTER 4: Catholic Social Teaching and the Environment — 81
Introduction — 81
Catholic teaching on environmental issues prior to *Laudato Si'* — 82
God the creator — 82
The relationship of man to animals and plants — 86
"Sustainable world" — 89
Marriage and the Family – the heart of civilisation — 90
Pope Francis reasserts Catholic social teaching — 94
The Book of Genesis — 95

The value of other living beings	97
Human sin and environmental degradation	98
Integral Ecology in *Laudato Si'*	101
Development of Catholic Doctrine	104
Laudato Si': some initial observations	108

CHAPTER 5: Laudato Si': a contribution to "the need for forthright and honest debate" — **111**

A word of caution	112
The preferential option for human life	114
Pope Francis and the Enlightenment	115
Some problematic aspects of *Laudato Si'*	117
Scientific authorities	118
Other environmental concerns	122
Bias in *Laudato Si'*?	123
The Church and scientific controversies	126
Pope Francis and deserts	127
Provocative language and questionable facts	130
"Fact" and who says so?	133
Global Inequality	140
Development assistance by country as percentage of Gross National Income 2013	141
High Rhetoric meets Doomsday	145
Climate Change	146

CHAPTER 6: In-Church Responses to Laudato Si' — **151**

Catholic Earthcare Australia	152
Political Misuse of the Encyclical	159
Caritas Australia	161
CAFOD UK	163
St Vincent de Paul Society Australia	165
Conclusion	169

CHAPTER 7: And Finally — **173**

FOREWORD

Like the encyclical from which it takes its name, this book ranges over a variety of modern challenges. Its chief virtue is the identification of what in my mind is one of the signal features of the modern world, namely, the reliance on immediate emotional reactions to situations. In short, a failure to recognise the importance of reason in critically examining assertions and theories. This is why Rev. Dr. Fleming recognises that the long term effectiveness of *Laudato Si'* rests on not just the moral authority of Pope Francis, but also on the reasonableness of its arguments. Of course, the arguments need to be persuasive, but they ought to be because of their reasonableness, not because they happen to be what seems to capture the popular, majority opinion. Thus, if *Laudato Si'* is to be persuasive, it cannot be because it contains what people want to hear or reinforces what they believe, it has to be because its arguments are based on what is true.

The truth is ever difficult to find, however, since there are many ways in which it can be obscured. If the truth *Laudato Si'* contains is to be grasped it is important firstly, to have a clear idea of the nature of an encyclical, secondly, to have some sense of its genesis and how it is continuous with Catholic tradition and teaching and thirdly, examine the scientific claims it supports. While in the past, it might have been expected that an educated class would understand the differences between encyclicals, *motu proprios,* apostolic exhortations, letters and other publications of the Pope, this cannot be assumed today. It is well that Rev. Dr. Fleming provides us with an overview of how the authority of an encyclical is to be understood. He also demonstrates how the encyclical is continuous with Catholic social teaching and questions the validity of the scientific claims. Care in particular should be taken to avoid partiality in relation to scientific claims, lest it obscures the truth.

Recognised in this book is the importance of history. This does not simply mean having a number of facts at one's disposal, but an understanding of how we have come to our present mode of thought. Crucially, the baneful influence of the Western Enlightenment is outlined, showing how Western thought became sundered from its roots in theology and hence an appreciation of how human beings depend on God. Human subjectivity became absolutized. Consequently, science, a human creation, becomes the ultimate source of knowledge and the yardstick by which human progress is to be measured. Little by little, science, so the proponents of modernity hold, will come to replace not only philosophy and ethics, but also religion. Faith in God is replaced by faith in science, and the technology that it spawns.

One of the great accomplishments of the Church is its championing of social justice. Indeed, some mistakenly believe that Christian faith is solely to be understood as social justice. Quite rightly, Rev. Dr. Fleming points out that the Church's commitment to social justice arises from its understanding of human beings as fallen creatures that Christ has redeemed and now calls to Himself. It is based in a Christian anthropology in which the fate of human beings is inextricably linked to their response to what God has ordained them to be. To heed the call to follow Jesus, human beings must love God with all their hearts, all their minds and all their strength, and their neighbour as themselves (Luke 10:27; Mark 12:30-31). Our neighbour is everyone, without distinction, since whatever we do for the least of our brothers and sisters we do to Jesus Himself (Matthew 25:40). Based on this understanding of human nature and the relationship that human beings have to each other and to God, Catholic social teaching is founded on three important principles: common good, subsidiarity and solidarity. It is these three principles, expressed in different ways, which guide the various Church documents elaborating the Church's teaching on social justice.

The Church's open support for the unity of faith and reason, is under attack in many parts of the world, including Australia, where arguments on many social issues are stifled, ridiculed and suppressed merely because they are putting a reasoned argument against what certain sections of the community hold. One such very public issue concerns the debate about so-called gay marriage and the attempt by its supporters to impose their views on Australian society. It takes courage to point out that the tactics employed by its supporters would not be out of place in the worst of totalitarian regimes. These typically begin with a denial of the right of anyone who disagrees to put their point of view, generally dismissing such a view as religious and medieval, even if no mention is made of religion or of the middle ages in any arguments put forward. It may be followed up with the threat of legal action for causing offence. There is complete disregard of rational argument in these tactics. It is the Church which steadfastly refuses to be intimidated by these threats and continues to argue for open, reasoned debate in the public square for all parties.

The origins of the Church's social teaching on the environment can be found in *Genesis,* where human beings are given stewardship over all other livings things, both plants and animals. Stewardship is to be understood as having responsibility for the welfare of all God's creatures, who belong to God, not to human beings. Dominion is not be understood as exploitation and abuse of the earth and its creatures, rather, governance on God's behalf, taking seriously the trust that has been placed in us to be conservators and not exploiters. Exploitation, however, is not confined to the other living things on the earth, but is a hallmark of how human beings treat each other. As the book points out, destruction of the family through undermining the institution of marriage, the annihilation of unborn children through abortion, the campaigning to embrace euthanasia and other actions damaging of human life are some of the ways in which human beings exploit and abuse each other. The environment consists not just of everything except human beings, but includes them and so environmental problems cannot be divorced from the challenges that an "anti-life" or "culture of death" ideology poses to human beings themselves: as *Laudato*

Si', in continuity with the Church's social teaching, points out, all things are connected.

Of course, Rev. Dr. Fleming does not agree with every aspect of the encyclical, taking issue, in particular, with the contention that there is consensus over climate change. He distinguishes carefully between the politics of climate science and the science itself. Scientists themselves, he observes, generally take a middle ground between those who are convinced deniers and those who are equally convinced affirmers of climate change. In his view, the encyclical is notably one-sided in supporting the reality of climate change, arguing that the science is not conclusive. Rev. Dr. Fleming contends that the language of the encyclical, in considering environmental issues, is inflammatory and partisan, and makes claims that are not all supported by evidence or for which there is contrary evidence. As a result, climate change affirmers take the encyclical as not only supporting their position on the environment, but also other, more radical proposals, such as, population control through contraception and abortion, which it decidedly does not. There is a danger that *Laudato Si'* will be used for political ends and Rev. Dr. Fleming presents evidence that this is precisely what has happened.

This is a provocative book, written by an author committed to the quest for truth through rational debate. It challenges its readers to engage critically with *Laudato Si'* and to read it carefully. It points out what it considers the flaws in the scientific account accepted in the encyclical and urges close examination of the scientific evidence. This is vital, if we are to provide a measured response to the many challenges that the encyclical addresses.

Professor Jānis (John) T. Ozolins, FHERDSA, FPESA, FACE
Australian Catholic University
Latvian Academy of Sciences

PREFACE

This book is intended as a strong defence of Catholic social teaching but also as a critical evaluation of the way in which that teaching is applied to our current situation. The essential thesis of this book may be summarised as follows:

- All Catholics must accept Catholic social teaching with religious submission of mind and will;
- *Laudato Si'* summarises Catholic social teaching where environmental and related issues are concerned;
- *Laudato Si'* applies Catholic social teaching to the current state of the environment as perceived by Pope Francis;
- The Pope's description of the state of the environment is open to debate;
- On climate change the Pope opts for and heavily promotes the "consensus" account of the science without referenced justification and does not refer to any other view;
- The Catholic Church lacks the competence to resolve continuing scientific debates;
- The debate over anthropogenic global warming is far from complete;
- *Laudato* Si', Pope Francis' Encyclical Letter *On the Care of Our Common Home*, repeatedly calls for a "forthright and honest debate" (nn. 16, 61, 135, 138, 188);
- The present author takes no "side" in the debate about climate change and wants the debate to continue, but in a context which provides for a much fairer hearing to be given to all sides to that debate;
- This book is intended as a contribution to that "forthright and honest debate" on environmental issues generally for which the Pope calls.

CHAPTER 1

Interpreting the Encyclical

On the 24th of May 2015 Pope Francis issued his "Encyclical Letter on Ecology and Climate". It is named *Laudato Si'* and subtitled *On Care For Our Common Home*.

But what is an "Encyclical Letter"?

In 1964 Pope Paul VI addressed this question directly in a General Audience held at Castel Gondolfo (the Pope's summer residence).[1] He told the people he was going to share a confidence with them. And that confidence was that he had

> … finally finished writing our first Encyclical Letter, which shall bear the date of the feast of the Transfiguration of Christ, tomorrow August 6, and the Latin text will begin with the words, *Ecclesiam Suam*, which will serve to identify it.[2]

Pope Paul VI then went on to discuss what an encyclical is.

> You know what an encyclical is. It is a literary work in the form of a letter sent by the Pope to the bishops of the whole world: an encyclical would mean a circular. It is a very ancient form of ecclesiastical correspondence. Its characteristic is that it denotes the communion of faith and charity existing between the various "churches", that is, among the various communities that make up the Church. In the early days even the heads of the main communities sent circular letters to fellow bishops and to all the faithful and therefore [these letters] were also called

---

[1] Pope Paul VI, *Udienza Generale*, Castel Gandolfo, Mercoledì, 5 Agosto 1964. The translation from the original Italian is my own.
[2] *Ibid.*

"catholicae," that is addressed to the entire Christian people. The historian Eusebius, in the fourth century, says that "ad universos Christi fideles dirigebantur" (ie they are directed to all Christians). (Hist. Eccl., V, 17).

In times closer to us encyclicals are addressed to a group of Bishops of a given region (Litterae), or to all the Bishops in communion with Rome (Epistulae), and sometimes also extended to all the faithful and even all men of good will depending on the content and purpose of the document. ***An Encyclical may be doctrinal or dogmatic, when it comes to the truth or errors related to faith; or it may be an exhortation*** if it tends to confirm in those who receive it the feelings and intentions of the Christian life, and to strengthen the bonds of discipline, of union, of fervour, which must be connected inside the church and support it in its spiritual mission.[3] [Emphasis added)

As Pope Paul VI puts it, the essential characteristics of the modern Encyclical Letter are:

1. The Letter is addressed not only to bishops but to "all men of good will";
2. The Encyclical may focus on matters of doctrine or dogma to clarify teaching and confirm in the hearts and minds of those who read it just what is authentic Christian teaching;
3. The Letter may be an exhortation understood as confirmation of the practical ways in which someone can lead an authentic Christian life.

While the writing of encyclicals goes back to the early centuries of the Church, the emergence of the encyclical as an authoritative teaching document in its present form dates from the papacy of Gregory XVI, Pope from 1831 to 1846. These days the term "encyclical" is almost exclusively used to describe certain papal documents which differ in their technical form from the ordinary

---

3 *Ibid.*

style of either Bulls[4] or Briefs.[5] They are letters explicitly addressed to the patriarchs, primates, archbishops, and bishops of the Universal Church in communion with the Apostolic See, and increasingly to "all men of good will". By exception, encyclicals are sometimes addressed only to the archbishops and bishops of a particular country. But we can safely say that encyclicals addressed to the bishops of the world are generally concerned with matters which affect the welfare of the Church at large.

Another way of answering the question as to the nature of an encyclical as it has evolved over the last 150 years is to see it as the favoured instrument to transmit the ordinary teaching of the Pope and, on comparatively rare occasions, to define infallibly a particular Church teaching. Typically:

> ...they set out an aspect of Catholic belief or teaching in a serious and substantive way, which is intended to be given significant weight by all members of the Catholic Church. In those (relatively rare) cases where an encyclical clarifies a central question of Christian belief or church doctrine on faith or morals (such as the forgiveness of sin through Christ's Passion and death, or the wrongfulness of corruption), its teaching is proposed expressly as true and "binding"; that is, as something offered for all Catholics to reflect on and freely accept as part of their faith in both belief and action.[6]

An example of infallible teaching in the modern era is that of Saint John Paul II's *Ordinatio sacerdotalis*, an Apostolic Letter in which he said:

> Wherefore, in order that all doubt may be removed regarding a matter of great importance, a matter which pertains to the

---

4 The most recent example of a Bull is Pope Francis' *Misericordiae vultus*, also titled Bull of Indiction of The Extraordinary Jubilee of Mercy.

5 A Brief is a less formal papal communication authenticated with a wax impression (now a red ink impression) of the Ring of the Fisherman.

6 M Casey and P Smith, "Catholic Social Teaching and Papal Encyclicals", http://www.ceosyd.catholic.edu.au/Parents/Religion/staff-faith-formation/Documents/01%20-%20Schools%20Resource%20-%20Laudato%20Si.pdf

Church's divine constitution itself, in virtue of my ministry of confirming the brethren (cf. Lk 22:32) I declare that the Church has no authority whatsoever to confer priestly ordination on women and that this judgment is to be definitively held by all the Church's faithful.[7]

Where encyclicals are concerned, another example of infallible teaching is this:

Therefore by the authority which Christ conferred upon Peter and his Successors, and in communion with the Bishops of the Catholic Church, *I confirm that the direct and voluntary killing of an innocent human being is always gravely immoral.*[8]

By and large, however, encyclicals overwhelmingly present the ordinary teaching of the Pope as he reflects upon the application of Catholic teaching to certain situations confronting the Church and the modern world. So an encyclical might include infallible definitions of teaching (rarely), clarification and elaboration of existing Church teaching (often), and in the case of Catholic social teaching how the doctrinal and dogmatic faith of the Church may be lived out in the modern world. Most encyclicals contain three components: doctrine, social questions, and pastoral questions.

How are we to interpret a particular encyclical?

When the Pope teaches in relation to faith and morals, and especially when he does so in an encyclical letter, his teaching is to be received by the faithful with religious submission of mind and will. The Fathers of the Second Vatican Council expressed it in this way in their *Dogmatic Constitution on the Church*.

This religious submission of mind and will must be shown in a special way to the authentic magisterium of the Roman Pontiff, even when he is not speaking ex cathedra; that is, it must be shown in such a way that his supreme magisterium is

---

7 *Ordinatio sacerdotalis,* 22 May 1994, n. 4,
8 *Evangelium vitae,* 25 March 1995, n. 57.

acknowledged with reverence, the judgments made by him are sincerely adhered to, according to his manifest mind and will. His mind and will in the matter may be known either from the character of the documents, from his frequent repetition of the same doctrine, or from his manner of speaking.[9]

That being said we still need to address a further question, and that is how we are to understand and interpret a particular document. For example, after the publication of the Apostolic Letter, *Ordinatio sacerdotalis*, some people questioned whether the Church was forever bound by the Holy Father's teaching "that the Church has no authority whatsoever to confer priestly ordination on women and that this judgment is to be definitively held by all the Church's faithful." A *dubium*[10] was then addressed by the Holy See. The *dubium* said this:

> Whether the teaching that the Church has no authority whatsoever to confer priestly ordination on women, which is presented in the Apostolic Letter *Ordinatio Sacerdotalis* to be held definitively, is to be understood as belonging to the deposit of faith.

The answer given was in the affirmative and then elaborated:

> This teaching requires definitive assent, since, founded on the written Word of God, and from the beginning constantly preserved and applied in the Tradition of the Church, it has been set forth infallibly by the ordinary and universal Magisterium (cf. Second Vatican Council, Dogmatic Constitution on the Church Lumen Gentium 25, 2). Thus, in the present circumstances, the Roman Pontiff, exercising his proper office of confirming the brethren (cf. Lk 22:32), has handed on this same teaching by a formal declaration, explicitly stating what is to be held always, everywhere, and by all, as belonging to the deposit of the faith.[11]

---

9 *Lumen gentium*, 21st November 1964, n. 25.
10 A *dubium* is a formally expressed doubt on a matter of faith or morals and submitted to the Holy See for adjudication.
11 Congregation for the Doctrine of the Faith, Responsum ad Propositum Dubium Concerning the Teaching Contained in "Ordinatio sacerdotalis", 28 October, 1995.

But to the more general question of interpreting an encyclical letter, the Australian Catholic Bishops' Conference has given this guidance:

> Overall then, the highest level of authority lies with encyclicals and Catholics are expected to accept their teachings as demanding our loyalty and eliciting our assent. ***However, not all encyclicals give us infallible teachings, and even in those that do, not everything taught in them is infallible. We are required to consider the content and context of each document, especially the intention of the Pope in issuing the document, and the way in which the bishops and the whole Church receive the teaching.*** Many encyclicals are important and weighty documents, but it was not the intention of the Pope to make a binding or infallible teaching in them. For example Benedict XVI's first encyclical, *Deus caritas est*, was a very important document but there is no new teaching being proclaimed as binding on the Church.[12] [Emphasis added]

So "the content and context" of an encyclical has to be considered. Over the last 110 years or so the encyclical has become an important instrument for the dissemination of Catholic social teaching. Catholic social teaching is derived from the Church's authority to teach on matters of faith and morals and in the case of a Papal Encyclical from the Pope's specific teaching authority. Even there, "the intention of the Pope in issuing the document" must also be considered together with "the way in which the bishops and the whole Church receive the teaching."

As remarked earlier, Pope Paul VI said that "an Encyclical may be doctrinal or dogmatic, when it comes to the truth or errors related to faith; or it may be an exhortation."

In his commentary on *Laudato Si'*, David Schütz considers this question: "What authority does *Laudato Si'* have?" His response to this question is both instructive and helpful.

> So what should an honest and open-hearted reader take from the

---

12 http://www.catholicaustralia.com.au/church-documents/types-of-documents

encyclical? The long term effectiveness of *Laudato Si'* will rest on three points:
1. The reasonableness of its arguments
2. The persuasiveness of those arguments, and
3. The moral authority of the author.

The Pope's moral authority does not trump the former two criteria ... The Galileo affair stands as a perpetual reminder: popes – even popes trained in science – are not infallible in any field outside faith and morality.[13]

Laudato Si' is an encyclical which is, in part, doctrinal, containing as it does Catholic social teaching, but provides no new dogmatic teaching, those teachings which must be considered to be part of the deposit of faith. The largest part of the document (about 80% of it) is an attempt to apply Catholic social teaching to the way the Pope sees the modern world and its environmental problems. Pope Francis has stated that *Laudato Si'* is a document which is now "added to the body of the Church's social teaching".[14] But this would apply to only about 20% of the Encyclical which will be discussed in greater detail in chapter 3. Chapters 4 and 5 will examine the 80% of the Encyclical which deals with the way the Pope sees the world and his proposed solutions to the ills that he perceives.

---

13 David Schütz, *A Guide to Reading Laudato Si?*, published with ecclesiastical approval by Mustard Seed Bookshop, Lidcombe, NSW, 7.
14 *Laudato Si'*, n. 15.

CHAPTER 2

The ecological crisis of the twentieth century

The 1960s and 1970s were a time of increasing public awareness of the major ecological problems confronting the world due to industrialization and the agrarian revolution. First remarked on by the Huxleys earlier in the century, Lynn White Junior was to take up the matter and, following the opinions of Julian Huxley pointed the finger at the problem – Christianity.

Problems identified in the 1970s included:

- pollution from factories and power plants;
- oil spills;
- raw sewage;
- toxic dumps;
- pesticides;
- freeway construction; and
- loss of wilderness and biodiversity.

The environmental problems evident in more developed countries were also being experienced in lesser developed countries as the developed countries moved in to exploit the natural resources of the undeveloped countries.

The successes of the environmental movements have been many. Today there is a far greater awareness of the dangers to the earth from human activity, and that awareness has led to effective action in many parts of the world. Equally, no one doubts that there is still much that needs to be done, especially when one adds in the impacts on the health and well-being of many peoples who live in poverty.

In 1962 air pollution was so bad that that it caused the death of

750 people in London (UK), while in 1965 air pollution in New York City caused the deaths of eighty people during a weather inversion that lasted four days. Also recognised were instances of lakes and rivers polluted by industrial wastes.

Added to those issues was the growing concern about the use of pesticides and fertilisers as competitive markets argued for the increasing need for high yield crops. Pesticides and fertilisers certainly enabled greater predictability where crop yields were concerned. But the side effects of these chemicals were becoming increasingly unacceptable to humans and to animals. All of this is now well recognised and there would appear to be near universal agreement that things had to change. And that included a radical rethink of philosophical and religious ideas prevalent in society at the time.

The Lynn White Intervention

On the 22nd of April 1970, Earth Day was celebrated for the first time in the United States. Many people consider this day to be the start of the modern environmental movement. However contentious that statement may be, for the purposes of this book this is a convenient point of departure. Soon after, in 1974, Lynn White Jr set the cat among the pigeons with his highly influential essay, "The Historical Roots of Our Ecological Crisis". In that essay he blamed Christianity for the ecological crisis facing humanity. His essay seemed to have greater impact than the earlier warnings of Julian Huxley on whose opinions it would seem that White drew.

Lynn White Jr (1907-1987) was an historian. He also had a Master's degree in theology from Union Theological Seminary. In this paper he argued that modern science is the outcome of Christian theology which places man at the centre of creation. This is so, he says, because of the creation stories in the book of Genesis, the first book of the Bible. White's interpretation of these creation stories are as inaccurate as they are idiosyncratic, but nevertheless they gained a certain currency among environmentalists who themselves knew little to nothing about how to understand the data from Holy Scripture. This is how White puts his case:

> While many of the world's mythologies provide stories of creation, Greco-Roman mythology was singularly incoherent in this respect. Like Aristotle, the intellectuals of the ancient West denied that the visible world had a beginning. Indeed, the idea of a beginning was impossible in the framework of their cyclical notion of time. In sharp contrast, Christianity inherited from Judaism not only a concept of time as non-repetitive and linear but also a striking story of creation. By gradual stages a loving and all- powerful God had created light and darkness, the heavenly bodies, the earth and all its plants, animals, birds, and fishes. Finally, God had created Adam and, as an afterthought, Eve to keep man from being lonely. Man named all the animals, thus establishing his dominance over them. God planned all of this explicitly for man's benefit and rule: no item in the physical creation had any purpose save to serve man's purposes. And, although man's body is made of clay, he is not simply part of nature: he is made in God's image.
>
> Especially in its Western form, Christianity is the most anthropocentric religion the world has seen. As early as the 2nd century both Tertullian and Saint Irenaeus of Lyons were insisting that when God shaped Adam he was foreshadowing the image of the incarnate Christ, the Second Adam. Man shares, in great measure, God's transcendence of nature. Christianity, in absolute contrast to ancient paganism and Asia's religions (except, perhaps, Zoroastrianism), not only established a dualism of man and nature but also insisted that it is God's will that man exploit nature for his proper ends.[15]

Since White's paper is not supported by references, it is difficult to know where to begin with all of this. It is certainly true that Genesis posits a beginning of the created order, a position re-emphasised

---

15 Lynn White Jr, "The Historical Roots of Our Ecological Crisis", *Ecology and Religion in History*, New York, Harper and Row, 1974.

by St John.[16] Thereafter there is not much with which to agree with White. Leaving aside White's curious description of Eve as God's "afterthought", there is no suggestion that man naming animals represents dominance. Indeed, Genesis describes all living creature as a gift from God to man to be man's helper and companion.

> Then the Lord God said, "It is not good that the man should be alone; I will make him a helper as his partner." So out of the ground the Lord God formed every animal of the field and every bird of the air, and brought them to the man to see what he would call them; and whatever the man called every living creature, that was its name. The man gave names to all cattle, and to the birds of the air, and to every animal of the field; but for the man there was not found a helper as his partner. So the Lord God caused a deep sleep to fall upon the man, and he slept; then he took one of his ribs and closed up its place with flesh. And the rib that the Lord God had taken from the man he made into a woman and brought her to the man.[17]

Far from being an anthropocentric account of creation, mankind (male and female) is presented in a cosmic context. Yes human beings are made in the image and likeness of God. But no, they are not to treat other living creatures as things to be abused, and unreasonably exploited. It is only when sin enters the world that we see animosity between the man and the woman, and between some animals and human beings. Specific Catholic teaching on the environment will be detailed and explained in the next chapter.

But let us explore the impact of the thinking of the Enlightenment on science, technology, politics and economics in relation to the subsequent impact of human activity on the environment.

---

16 John 1:1.
17 Genesis 2: 18-22.

The Enlightenment, Secularists, and the Environment

The contemporary concern about environmental issues is generally shared within the community, within states, and among the community of states. The secular thinking responsible for the mistreatment of the natural world is rarely recognized, with the Catholic Church often being attacked as the enemy of modernity. The very kind of thinking that led, in turn, to both the scientific revolution and with it the industrial and technological revolutions, is now being appealed to as the basis for finding solutions to the very problems created from the natural consequences of that thinking. But the problems connected with this thinking will not be resolved by more of the same, with man's arrogance in displacing God as Lord of the earth leading to unprecedented levels of environmental degradation.

That this is the case is rarely acknowledged by many contributors to the debate over environmental degradation. Lynn White Jr, however, partially admits the problem but has no solution to it.

> What we do about ecology depends on our ideas of the man-nature relationship. More science and more technology are not going to get us out of the present ecologic crisis until we find a new religion, or rethink our old one. The beatniks, who are the basic revolutionaries of our time, show a sound instinct in their affinity for Zen Buddhism, which conceives of the man-nature relationship as very nearly the mirror image of the Christian view. Zen, however, is as deeply conditioned by Asian history as Christianity is by the experience of the West, and I am dubious of its viability among us.[18]

It has to be admitted that some of the worst excesses of Enlightenment thinking have infected the thinking of many Christians in their quest for "relevance" such that the basis of the published ideas of many Christian theologians seem indistinguishable from secularists. In the Catholic context, Hans Kung and Paul Collins are representative

---

18 Lynn White Jr, *op. cit.*, 5.

of precisely what happens when Catholic scholars embrace almost uncritically the normative moral positions of secularists.

For the rest of this chapter any reference to humanism is a reference to *secular* humanism, that humanism which disconnects itself from the idea of God, seeing God and "religion" as necessarily irrational, anti-progress, and an exercise in repressive superstition. Let the secular humanists define humanism:

> HUMANISM is a way of thinking and living that aims to bring out the best in people, so that all people may have the best life. Humanists reject supernatural and authoritarian beliefs. They consider that we must take responsibility for our own lives, and show care and compassion towards family and community. Humanists consider that conserving the habitats of all species is central to a sustainable future. The Humanist lifestyle emphasises reasoned enquiry and dialogue, freedom, responsibility, and the need for tolerance and cooperation.[19]

Magister mundi sum! – I am the master of the universe!

Secular humanism, conceived and born in the challenge that science brought to religion in the seventeenth century, permeated Western culture and thinking over the succeeding centuries. Francis Bacon's pronouncement that "human knowledge and human power meet in one; for where the cause is not known, the effect cannot be produced"[20] – that knowledge is power – is fundamental to humanist thinking. For Bacon, the "fifth essence" is not to be found in Aristotle's "fantastic heaven" but in the "Eupolis" or "good city."[21] Natural philosophy had been corrupted by the search for the "final cause" as the end for which all things exist. The final cause, for Bacon, is:

---

19 *Australian Humanist*, No. 117, Autumn 2015.
20 Francis Bacon, *Great Instauration*, bk. 1, para 99 cited in Hiram Caton, *The Politics of Progress*, Gainesville: University of Florida Press, 1988, 41
21 Howard B. White, "Francis Bacon", in *History of Political Philosophy*, eds. Leo Strauss and Joseph Cropsey, Chicago and London, The University of Chicago Press, 1987, 366.

the things which man makes, and the best thing that man can make is the best commonwealth, the commonwealth of the New Atlantis, the happy land, the land of all earthly things worthiest of knowledge.[22]

This "elevation of the man-made, the artificial, over the natural" is the basis of Bacon's materialist philosophy. The ideal state is realised through the conquest of nature by mankind in order to achieve the relief of those agonies which afflict the human condition. Some "Baconian enthusiasts during the Commonwealth period understood the new philosophy as benevolent and utopian", yet, says Hiram Caton, "the linkage between power and goodness is riddled with difficulties, the most apparent being that the link is wholly contingent."[23] Bacon's reply that everything can be abused is "hardly adequate", observes Caton, "since his proposal for unprecedented increase in human power opened the door to unheard-of abuses".[24]

Caton draws attention to "a second set of statements on the ends of power, hardly noticed today" in which Bacon "dropped the unconvincing appeal to charity and related power to the end it typically seeks in human life, greatness."[25]

> My purpose is to try whether I cannot in fact lay more firmly the foundations, and extend more widely the limits, of the power and greatness of man.[26]

After this assessment of what he regards as legitimate human ambition Bacon unveils "in a skyrocket of sparkling prose the grandest design of all: 'But if a man endeavour to establish and extend the power and dominion of the human race itself over the universe, his ambition (if ambition it can be called) is without

---

22 *Ibid.*, 367.
23 Hiram Caton, *The Politics of Progress*, Gainesville, University of Florida Press, 1988, 42.
24 *Ibid.*, 43.
25 *Ibid.*, 44.
26 Francis Bacon, *Great Instauration*, bk. 1, para 116 cited in Hiram Caton, *op. cit.*, 44.

doubt both a more wholesome thing and a more noble than the other two.'"[27] Caton goes on to remark that ambition "in such Gargantuan excess" needs another name. "Hope of dominion over the universe bugles an assault on Mount Olympus to overthrow the kingdom of God."[28]

The link between science and life is power and greatness in Baconian philosophy, a link "forged in the first instance by his [Bacon's] methodological requirement of unified science, which demands that all inquiry search for the effective truth of efficient causes."[29]

Here, as early as the seventeenth century, we see the revolution in human thinking which, acknowledging the Greek philosopher Protagoras, sees man as the measure of all things[30] together with a confidence that human beings will always, or nearly always, harness the power of science and technology for good rather than ill. The rest of the material world is the playground in which man can assert his god-like superiority over the created order through the gradual increase in knowledge which in turn unleashes power. God decreases as man increases. And set free from the "shackles" of religion, arbitrarily defined as irrational by all Enlightenment *bien pensants*, man will put right what this "god" got wrong.

---

27 Francis Bacon, *Great Instauration*, bk. 1, para 129, cited in Hiram Caton, *ibid.*, 44.
28 Hiram Caton, *Ibid.*
29 *Ibid.*
30 Based on a statement by the fifth century BC Greek philosopher Protagoras. It is usually interpreted to mean that the individual human being, rather than a god or an unchanging moral law, is the ultimate source of value.

But as long ago as 1978, David Ehrenfeld[31] drew attention to the arrogance of humanism. For Ehrenfeld, this "arrogance of the humanist faith in our abilities was nurtured by the late Renaissance triumphs of science and technology working in tandem."[32] Ehrenfeld argues that the source of the problem is the doctrine of final causes, a doctrine which he says has "flourished since the rise of the sciences in the late sixteenth and seventeenth centuries." As Ehrenfeld formulates it, this doctrine asserts "that the features of the natural world ... have all been arranged by God for certain ends, primarily the benefit of humanity."[33] The transition to secular humanism could occur in steps. Humans are made in God's image.

---

31 David Ehrenfeld is an American professor of biology at Rutgers University and is the author of over a dozen publications, including *The Arrogance of Humanism* (1978), *Becoming Good Ancestors: How We Balance Nature, Community, and Technology* (2009), and *Swimming Lessons: Keeping Afloat in the Age of Technology* (2012). He is often described as one of the forerunners of twentieth-century conservation biology. Ehrenfeld's work primarily deals with the inter-related topics of biodiversity, conservation, and sustainability. He is also the founding editor of Conservation Biology, a peer-reviewed scientific journal that deals with conserving the biodiversity of Earth, and has written for various magazines and newspapers including the New York Times, the Los Angeles Times, and Harper's Magazine.

32 David Ehrenfeld, *The Arrogance of Humanism*, (New York: Oxford University Press, 1978), 12.

33 *Ibid.*, 7. This formulation appears to me to be inadequate. A simple explanation of the doctrine may be found in Ralph McInerny, *First Glance At St. Thomas Aquinas*, Notre Dame, University of Notre Dame Press, 1990, 82-89. Ehrenfeld appears to have confused the doctrine of ultimate causality with the principle of finality. The doctrine of ultimate causality or first cause states that God "is the unique first cause of all realities that exist outside himself." Ronda Chervin and Eugene Kevane, *Love of Wisdom*, San Francisco, Ignatius Press, 1988, 186. The principle of finality "states that every efficient cause, namely, everything that acts for doing or making something, has a purpose in doing so ... An archer has a conscious purpose, and the arrows fly purposefully. But the arrow in itself has no conscious awareness of the purpose toward which it has been aimed. Another way of stating this is that everything exists for some good, namely, the purpose or end in view." *Ibid.*, 134. The ultimate or final end of man is God who is also the first cause. [Unde dicitur quod finis est causa causarum] *Cf Summa Theologiae*, 1a, q. 5, a. 2, ad. 1; 1a, q. 105, a. 5. Some of the "features of the natural world", to which Ehrenfeld refers as having "been arranged by God", are due to human action. Other relevant references include *1 Contra Gentes, 13*; *Summa Theologiae*, 1a-2ae, q. 109, a. 6; 1a-2ae, q. 70, ad. 2.

God could then be retired and eventually abandoned. What remains is the god-like humanity of the Christian tradition for whose benefit the world exists. Ehrenfeld's account of the way the transition to humanism occurred is problematic. What is not in doubt is that such a transition has occurred, and that humanist thinking has pervaded Western culture, including many of the contemporary Christian theologies, as well as those philosophies that appeared as a bridge between religion and political theory.

For example Karl Rahner, the influential Jesuit theologian, reflects much of the Baconian humanist account in his essay, "The Experiment with Man". For Rahner "man has become what, according to the Christian understanding he is: the free being who has been handed over to himself." Triumphantly Rahner proclaims that "as a real partner of the God of the 'other world' man can and must stand over against this world as its lord." [34]

Rahner's triumphant anthropomorphism is more than matched by the breathtakingly materialistic and reductionist account which exemplifies the secular humanist cause as exemplified here by leading secular humanist Ian Bryce:

> Research in neuroscience has shown how brains work at a basic level, and is yielding clues into the human mind, behavior, and emotion. Thought, reasoning, freewill, dreams, conscience, and consciousness are revealed as internal functions of the brain. Astral travel, near death experiences, prayer, psychic abilities, and telekinesis must be illusory, as physics provides no means of communication apart from the familiar senses. Thus such pseudoscience assertions can be dismissed as nonsense.[35]

Moreover, Ian Bryce has no inhibitions in revealing his intense contempt for Catholics and their rights as human beings:

> Figures such as Locke, Rousseau, Diderot, Condorcet, Kant,

---

34 Karl Rahner, "The Experiment with Man", in *On Moral Medicine*, eds. S. E. Lammers & A. Verhey, Grand Rapids, Michigan, William B. Eerdmans Co., 1987, 232.

35 Ian Bryce, "Science, Enlightenment and Humanism", *Australian Humanist*, No. 117 Autumn 2015, 5.

> Hume, Paine, Jefferson, Franklin and Leibniz are well known for their contributions both to science and hence to society. *A favourite is Carvalho, the Marquis of Pombal, who in 1761, arranged the trial and execution of Malagrida, the leading Catholic in Portugal, and hence brought the Inquisition to sudden stop. Is execution justified if it curbs a tyrannical regime?*[36] [Emphasis added]

How interesting that Bryce should refer to the little known Malagrida affair among the list of democratic enlightenment figures like Locke, Hume and Kant. The Malgrida affair is instructive for what it reveals about what Bryce the humanist thinks should be valued.

In his representation of the Malgrida affair to which he refers approvingly, Bryce reveals the authoritarian aspect of so-called "democratic enlightenment" where those who think they know best because they are "enlightened" have no problem in imposing their ideas even on unwilling populations. So Bryce takes side with the autocratic Carvalho, the Marquis of Pombal and a leading figure in the Portuguese Enlightenment. The reality is that Carvalho used the Inquisition (of which he fundamentally disapproved) to deal with the Jesuit Father Malagrida who he wanted eliminated simply because he was a Jesuit, and as part of Carvalho's programme to rid Portugal of Jesuit influence.

> Without proof, Malagrida was declared guilty of high treason, but, being a priest, he could not be executed without the consent of the Inquisition. Meanwhile the officials of the Inquisition, who were friendly towards Malagrida, were replaced by tools of Pombal, who condemned him as a heretic and visionary, whereupon he was strangled at an auto-da-fé, and his body burnt. The accusation of heresy is based on two visionary treatises which he is said to have written while in prison. His authorship of these treatises has never been proved, and they contain such ridiculous statements that, if he wrote them, he

---

36 *Ibid.*, 4.

must previously have lost his reason in the horrors of his two and a half years' imprisonment. That he was not guilty of any conspiracy against the king is admitted even by the enemies of the Jesuits. A monument in his honour was erected in 1887 in the parochial church of Menaggio.[37]

There is no doubt that Carvalho achieved many fine things economically for Portugal, and was a forceful and effective leader after the 1755 earthquake. At the same time, however, he used his "enlightenment" to entrench his own autocratic political power at the expense of individual liberty. Moreover, his intensification of the economic exploitation of the Portuguese colonies allowed him opportunity to enrich himself at the expense of others.[38] It is true that Carvalho ended the Inquisition but only after he had staffed it with his own men and hypocritically used it to get rid of Malagrida. Bryce's rhetorical question, "Is execution justified if it curbs a tyrannical regime?" reveals the unpleasant authoritarian, unjust, and anti-religion side of the secular humanist perspective which is all too prevalent among environmental political activists today.

And Oxford University chemist, Peter Atkins, makes this astonishing claim for the all-seeing and all-knowing capacity of science:

> *The scientific method can shed light on every and any concept,* even those that have troubled humans since the earliest stirrings of consciousness and continue to do so today. It can elucidate love, hope, and charity. It can elucidate those great inspirations to human achievement, the seven deadly sins of pride, envy, anger greed, sloth, gluttony, and lust.[39] [Emphasis added]

Such unqualified universal claims are indicative of a secular hubris that knows no bounds.

---

37 http://www.newadvent.org/cathen/09565c.htm
38 Kenneth Maxwell, *Pombal, Paradox of the Enlightenment*, Cambridge, Cambridge University Press, 1995, 83, 91-108, 160-62.
39 Peter Atkins, *On Being: A scientist's exploration of the great questions of existence*, Oxford University Press, 2011, Prologue.

The impact of secular political philosophy on theology

Leslie Stevenson traced the development of political philosophy from Hegel to Marx. Hegel's theory of historical development suggests mental or spiritual laws which allow cultures to develop, since cultures and nations have "a kind of personality" of their own.[40] The whole of reality is identified with God, what Stevenson refers to as "a pantheist rather than a Christian concept of God". For Hegel, human history is to be interpreted as "the progressive self-realization of this absolute spirit."[41] In the application of Hegel's ideas to politics his followers divided between those who believed that "the process of historical development automatically led to the best possible results", i.e. the Prussian State, and those who thought that the nation states of their time were far from perfect, it being the duty of all to work towards the "development of the next stage of human history."[42]

Feuerbach departed from Hegel on the matter of God progressively realizing Himself in history. For Feuerbach "the ideas of religion are produced by men" who are "dissatisfied or 'alienated' in their practical life". Stevenson summarises Feuerbach thus:

> Metaphysics is just 'esoteric psychology', the expression of feelings within themselves rather than truths about the universe. Religion is the expression of alienation, from which men must be freed by realizing their purely human destiny in this world.[43]

Stevenson concludes that Feuerbach is "one of the most important sources of humanist thought."[44] His influence on Marx was profound. Freed from the shackles of religion, the "opiate" of the masses, men and women could work towards the establishment

---

40 Leslie Stevenson, *Seven Theories of Human Nature*, Oxford, Oxford University Press, 1974, 46.
41 *Ibid.*, 47.
42 *Ibid.*
43 *Ibid.*
44 *Ibid.*

of the successive stages of human history, the bourgeois capitalist phase, the socialist revolution, and finally to communism when the state would wither away, there being no longer any need for police or external coercion. Thus would human beings reach the humanist utopia, through the inexorable forces of human history in which human beings take matters into their own hands, just as Bacon much earlier said they should. No longer would human beings be alienated from society because all property would be held in common. Human effort would have wholly benign consequences, even though blood would need to be shed to reach this end point of human history.

In the twentieth century, the influences of Feuerbach, Hegel, Marx and Darwin have been felt in new Christian theologies which have been developed in response to the new human situation. The conduit of these influences has been the revival of hermeneutics[45] and its central place in theological discussion. How can Christianity be interpreted in the light of the crises facing contemporary human cultures, and in such a way that theology is comprehensible to modern humankind? The powerful intellectual currents running through contemporary human societies include, according to Van A. Harvey, contributions from sociology, anthropology, history, literary criticism, philosophy, and religion. However:

> underlying all these currents is the assumption that human consciousness is situated in history and cannot transcend it – an assumption that raises important questions concerning the role of cultural conditioning in any understanding.[46]

Hans Georg Gadamer resumed the meditation of Martin Heidegger upon the crisis indicated by Nietzsche and formulated the issue as follows: "since all normative traditions have been rendered

---

45 For an interesting discussion on the way in which hermeneutics has developed over the centuries, see "hermeneutics" in *Stanford Encyclopedia of Philosophy*, 9th November 2005, at http://plato.stanford.edu/entries/hermeneutics/

46 Van A. Harvey, "Hermeneutics", *The Encyclopedia of Religion Volume 6*, editor in chief Eliade Mircea, New York, Macmillan Publishing Company, 1987, 280.

radically questionable, hermeneutics ... has become a universal issue."[47] According to Lawrence the responses to Gadamer were "fragmenting on the one hand, and totalizing on the other." These responses, he said, "bore the earmarks of that sort of interpretation that Marx in his famous eleventh thesis on Feuerbach, said needed to be supplanted by practice."[48] The consequence of this was the development of "liberation theologies" and other political theologies. These theologies (if indeed that is what they are), were developed in the European theological schools by such leading proponents as the German Catholic J.B. Metz and the German Protestant Jürgen Moltmann, and in the *communidades de base* [basic communities] of Latin America by Gustavo Gutiérrez (Peru), Juan Segundo (Uruguay), José Miguez-Bonino (Argentina), Jon Sobrino (El Salvador), and Leonardo Boff (Brazil). Both styles are meant, according to Lawrence, "to come to terms with the universal hermeneutic problem as portrayed by Nietzsche, Heidegger, Gadamer, and Paul Ricoeur. But it is no less evident that they mean to follow Marx's imperative of changing, rather than merely interpreting, history."[49]

Building on the theory of evolution some new theologies see humankind as the "spearhead of evolution", and it is this, "from a phenomenological point of view ... that gives him his status of dignity and superiority over his natural surroundings."[50] Denis Edwards, following Karl Rahner, sees "the human person ... as the cosmos come to consciousness of itself."[51] Edwards acknowledges that this line of thought can be traced back to Julian Huxley via

---

47 Frederick G. Lawrence, "Political Theology", *The Encyclopedia of Religion Volume 11*, 405.
48 *Ibid.*
49 *Ibid.*
50 N.M. Wildiers, *An Introduction to Teilhard de Chardin*, London, Fontana Books, 1968, 81.
51 Denis Edwards, *Jesus and the Cosmos*, Mahwah, New Jersey, Paulist Press, 1991, 40.

Teilhard de Chardin.[52] This new anthropocentrism, says Edwards, "differs from traditional anthropocentrism because it is profoundly relational. It views human beings as intimately related to the Earth and as 'companions' to every other creature."[53] Edwards does not develop the practical implications of this sort of theology, but he does note that some, "unjustly" in his view, "criticised the evolutionary work of Teilhard and Rahner as being naively optimistic."[54]

But the criticism of naïve optimism is, in fact, more than justified. For example, Wildiers, in his introduction to de Chardin's work, contrasts the "gradual process of degradation (entropy) and disintegration" of matter, which radically affects the animal and plant kingdoms, with the evolutionary history of mankind. He notes the extinction of many species and that there is no sign of new species appearing.

> Man, on the other hand, moves steadily onward and upward. As a species he shows no trace of any loss of vital energy. Numerically, he is still on the increase. His mental activity and his urge to expand are always intensifying. Of all sources of energy in the world he is the most dynamic ... Through man as the highest and central phenomenon that world evolution has produced that same evolution is bound to follow its ascending course ... Within the framework of the fundamental laws of nature man is the architect of to-morrow's world.[55]

This "anthropocentrism" so enthusiastically embraced by the "new theologies" is not really theology at all. Theology is the science of the study of God. Anthropology is the study of man. Accepting the Enlightenment project, that man's capacity to rule this world is almost limitless, we see how God is pushed to the margins, leaving man to control his destiny on his own terms. Walter Cardinal

---

52 *Ibid.* De Chardin said: "Man discovers that he is nothing else than evolution become conscious of itself, to borrow Julian Huxley's striking expression." Pierre Teilhard de Chardin, *The Phenomenon of Man*, London, Fontana Books, 1969, 243.
53 Denis Edwards, *op. cit.*, 43.
54 *Ibid.*, 15.
55 N.M. Wildiers, *op. cit.*, 82

Kasper, influential Catholic theologian, in tune with Karl Rahner, exhibited his own commitment to the Enlightenment cause when he said this:

> *The God who is enthroned over the world and history as a changeless being is an offence to man.* One must deny him for man's sake, because he claims for himself the dignity and honour that belong by right to man... We must resist this God, however, not only for man's sake, but also for God's sake. He is not the true God at all, but rather a wretched idol. For a God who is only alongside of and above history, who is not himself history, is a finite God. If we call such a being God, then for the sake of the Absolute we must become absolute atheists. Such a God springs from a rigid worldview; he is the guarantor of the status quo and the enemy of the new.[56] [Emphasis added]

While the Kasper approach to God and history was officially condemned by the Congregation for the Doctrine of the Faith in the 1984 instruction on liberation theology, it still persists to this day in the intellectual positions adopted by many Catholic theologians. The CDF taught:

> History thus becomes a central notion. It will be affirmed that God Himself makes history. It will be added that there is only one history, one in which the distinction between the history of salvation and profane history is no longer necessary. To maintain the distinction would be to fall into "dualism". Affirmations such as this reflect historicist immanentism. [...]
>
> Along these lines, some go so far as to identify God Himself with history and to define faith as "fidelity to history", which means adhering to a political policy which is suited to the growth of humanity, conceived as a purely temporal messianism.
>
> As a consequence, faith, hope, and charity are given a new

---

56 "Gott in der Geschicte", *Gott heute: 15 Beiträge zur Gottesfrage*, (Mainz, 1967) Translation from "The New Pastoral Approach of Cardinal Kasper to the divorced and 'remarried'", 12 April 2014, *Documentation Information Catholiques Internationales*, http://www.dici.org/en/documents/thenew-pastoral-approach-of-cardinal-kasper-to-the-divorced-and-remarried/

content: they become "fidelity to history", "confidence in the future", and "option for the poor." This is tantamount to saying they have been emptied of their theological reality.⁵⁷

Ehrenfeld identifies this "absolute faith in our ability to control our own destiny" as "a dangerous fallacy".⁵⁸ Humanism is based upon the "principal assumption, which embraces all of our dealings with the environment, and some other issues as well".⁵⁹

> It says: *All problems are soluble.* In order to make its connection with humanism clear, just add the two words that are implicit; it becomes: *All problems are soluble by people.*⁶⁰

To this principal assumption Ehrenfeld adds five others:

1. Many problems are soluble by technology.
2. Those problems that are not soluble by technology, or by technology alone, have solutions in the social worlds (of politics, economics etc.)

---

57 Voice of the Family, "Analysis of the Instrumentum Laboris of the Ordinary Synod on the Family", 2015, http://voiceofthefamily.com/wp-content/uploads/2015/10/Analysis-of-the-IL-of-the-Ordinary-Synod.pdf

58 David Ehrenfeld, *op. cit.*, 9-10. Ehrenfeld does not argue that there is nothing of value in humanism. "Belief in the nobility and value of humankind and a reasonable respect for our achievements and competences are also in humanism, and only a misanthrope would reject this aspect of it." *Ibid.*,10. *Cf* "The better parts of humanism are not in question here; when the inappropriate religious elements have been removed, humanism will become what it ought to be, a gentle and decent philosophy and a trustworthy guide to non-destructive human behaviour." *Ibid.*, 5. H. Tristram Engelhardt, Jr attempts a construction of bioethics on secular humanistic presumptions. In doing this he makes a distinction between Secular Humanism of the kind which Ehrenfeld describes as arrogant, and secular humanism as a "cluster of philosophical, philological, moral, and literary ideas, images, and commitments, which have been associated with the historical phenomenon of humanism in dissociation from particular religious or ideological commitments." H. Tristram Engelhardt, Jr., *Bioethics and Secular Humanism*, London: SCM Press, 3. Nevertheless, Engelhardt makes his own profession of faith: "Moreover, though faith in reason is largely lost, I have not lost the Faith." *Ibid.*, xvii.

59 David Ehrenfeld, *op. cit.*, 16.

60 *Ibid.*

3. When the chips are down, we will apply ourselves and work together for a solution before it is too late.
4. Some resources are infinite; all finite or limited resources have substitutes.
5. Human civilisation will survive.[61]

Put another way, man is free to reorder the creation in any way that suits his perceived interests at the time. Any problems are either foreseeable or not foreseeable. But in any case they will be able to be "fixed" by the same sorts of philosophical, scientific, technological, economic, and political reasoning that got us into the mess in the first place.

Ehrenfeld is concerned about environmental issues and what he sees as the impact of humanist thinking on industrial societies as a root cause of our present environmental woes. His essential criticisms of humanism have a broader scope and certainly embrace optimistic ethical accounts of scientific achievements in the medical sciences.[62]

Ehrenfeld's summary of the present mentality suggests that both "humanism and modern society have opted, albeit unconsciously, for the assumptions of human power." The choice, he says, "was understandable – the assumptions have long seemed, superficially, to work, and they certainly have been (and still are) gratifying to the ego."[63] According to Ehrenfeld's analysis, our faith in what science can do is often unreal and misplaced. He gives many examples of this. One will suffice here.

Referring to a paper by Professor E.C. Zeeman, entitled "A

---

61 *Ibid.*, 17
62 Joseph Fletcher, the American medical ethicist, reflects the humanist ethical account when he says: "A 'test tube baby', although conceived and gestated ex corpo, would nonetheless be humanly reproduced and of human value. A baby made artificially, by deliberate and careful contrivance, would be more human than one resulting from sexual roulette - the reproductive mode of the subhuman species." Joseph Fletcher, "Indications of Humanhood: A Tentative Profile of Man", *Hastings Center Report*, 2:5: November, 1972.
63 David Ehrenfeld, *op. cit.*, 21.

Model For Institutional Disturbances", which appeared in the May 1976 issue of the *British Journal of Mathematical and Statistical Psychology*, he discusses the capacity of science to measure and predict the future. Zeeman argued that his advanced mathematical theory, applied to "catastrophe theory", would enable them to be able to predict, and therefore to prevent, prison riots.[64] Ehrenfeld regards this paper as "a veneer of unusually sophisticated mathematics applied over the all-too-common base of ignorance and contempt of fellow human beings in trouble."[65] What "catastrophe theory" does, he says, is to simplify a very complex set of events to two control variables. Quoting Jonathan Rosenhead in the *New Scientist* he notes that a "social system of quite remarkable complexity must be simplified almost out of existence."

> This is reminiscent of Kraus's warning about simplified model systems, and about the tripartite absurdity of long-range predictions: we cannot know and gather in advance all the information that will be relevant, we cannot know what questions to ask of it, and if we did we could not make errorless deductions from what we know.[66]

Ehrenfeld insists that the appropriate word to describe the sorts of humanist themes to which he has referred is *arrogance*.

> The claims of predicting the unpredictable and of knowing the unknowable, the absolute faith in procedures whose end-results can never be comprehended – these things appear repeatedly. We are dealing with the same phenomenon in psychological testing, in cliometrics, and in the psychosocial applications of catastrophe theory. Where does such arrogance come from? I can only think of the mass persistence into adult life of that state of mind known as "magical thinking."[67]

---

64 *Ibid.*, 36-37.
65 *Ibid.*, 76.
66 *Ibid.*. Eric Kraus is a meteorologist. The reference here is to his "The Unpredictable Environment", *New Scientist*, 63, 1974, 649-52.
67 David Ehrenfeld, *op. cit.*, 77-78.

A persistent theme in humanist thinking is its assumption of responsibility for everyone on earth, which presumes that what is best for everyone can be known. Not all humanists take such a sanguine view of human nature. Paul Kurtz has observed that:

> humanists are not immune to moral corruption either. I have learned from direct personal experience in humanist organizations that even so-called 'humanists' will at times use mendacious means to achieve their goals, and that they are as prone to vanity, jealousy, vindictiveness, and other foibles as other human beings. was dismayed to discover that some so-called 'humanitarians' and 'philanthropists' make contributions or are devoted to a cause not for the good they will achieve, but for personal power and acclaim.[68]

Two years after Kurtz noted that humanists were as likely to be morally corrupt as anyone else, the leading US humanist Lyle L. Simpson was quoting, with approval, Abraham Maslow, "founder of humanistic psychology and the 1967 recipient of the Humanist of the Year Award". It was as if Kurtz's penetrating observation was not sufficient to penetrate Simpson's hubristic confidence in human nature set free from "religion". Having rehearsed Maslow's basic steps towards self-actualisation, Simpson observes that "no country has reached this utopian state of actualisation, and very few persons even understand its meaning."[69] Not deterred by this unfortunate state of affairs, Simpson asserts that it is "the effort of organized humanism" to "bring about this understanding and fulfilment."[70] Presumably humanists are among the few who have reached Maslow's utopia and are in a position to act as guides and instructors for the rest of us. "Although few may attain this level," says Simpson, then, as Maslow himself reminds us:

---

68 Paul Kurtz, "Does Humanism Have an Ethic of Responsibility", in *Humanist Ethics*, ed. Morris B. Storer, (New York: Prometheus Books, 1980), 23.
69 Lyle L. Simpson, "The State Of Humanism", *The Humanist*, January/February 1982, 38.
70 *Ibid.*, 39

> The purpose of living is to become fully functioning human beings in resonance with the universe. For this reason, our principal concern is the quality of life for each person on earth.[71]

The assumption of responsibility for the whole world is evident in the Humanist Manifestos I and II, and even more explicitly in the subsequent Humanist Manifesto III[72]:

> The ultimate goal should be the fulfilment of the potential for growth in each human personality - not for the favored few, *but for all of humankind* ... We affirm a set of common principles that can serve as a basis for united action - positive principles relevant to the present human condition. *They are a design for a secular society on a planetary scale.*[73] [My emphases]

Corliss Lamont also accepts that the "supreme ethical end for Humanists is the worldwide Community Good, that is, the welfare, progress, and happiness *of the entire human race*".[74] Utopia must be built in this world or not at all, while "reason and scientific method" are to be relied on for the "working out [of] ethical decisions".[75]

John Dunphy, too, believes that "all are responsible for all", a principle which he sees as the corollary of the proposition that God does not exist.

> If God has failed in his role as cosmic policeman and if Christianity has failed to uphold the dignity of humankind and to protect the inalienable rights of all – and who can argue with either hypothesis – then a viable alternative to both must be

---

71 *Ibid.*
72 http://americanhumanist.org/humanism/humanist_manifesto_iii
73 Lloyd L. Morain, "Humanist Manifesto II A Time For Reconsideration?", *The Humanist*, September/October 1980, 5.
74 Corliss Lamont, "The Affirmative Ethics of Humanism", *The Humanist*, March/April 1980, 6.
75 *Ibid.*, 53.

sought. That alternative is humanism.[76]

In his prize-winning essay Troy A. Jacobs depicts recombinant genetics as "an exercise in biology and humanism", and although it "can be used to change the evolution of a cell ... it can also change the attitude of the status quo religions" "because it is so intensely human".[77] Jacobs asserts both his faith in science and in a humanism which enables rational and scientific programmes to improve the world.

> Molecular biology, recombinant genetics, and other biological advances have shown that humans can tinker with the cell, with life, as if it were a machine. *Humans, too, are machines.* Still many years off, *the New Biology offers the possibility of cloning and revitalizing humankind.*[78] [My emphases]

As if to offer the world some protection from this tinkering, Jacobs suggests that "the United Nations should start work on humanity's basic biological rights." His proposal is that:

> 1. Except in unusual cases, all human beings be allowed to be born free of a genetically engineered genome.[79]

It is interesting to note Jacobs' notion of human rights. Instead of the inherent and inalienable right to life and other rights we merely have a humanist "permission" to be born free of a genetically engineered genome, unless of course one fits into the category of an "unusual case". In which case:

> 2. Whenever recombinant genetics can knowingly (by experience on lower hominoids) improve the life and survivability of a developing zygote, applied genetics technology may be used.[80]

---

76 John J. Dunphy, "A Religion For A New Age", *The Humanist*, January/February 1983, 25-26. Dunphy's article was one of 32 prize winning articles chosen from 300 entries in the North American Essay Contest run by *The Humanist*.
77 Troy A. Jacobs, "The Role Of Recombinant Genetics In Humanism", *The Humanist*, January/February 1983, 20.
78 *Ibid.*
79 *Ibid.*, 19.
80 *Ibid.*

It is not clear who or what are to be regarded as the "lower hominoids". It is also noteworthy that genetics technology may be done merely to "improve life". This catch-all term need not necessarily carry only a benign meaning.

Today nothing much has changed. Humanists still proceed on the basis of faith in human beings and their abilities in science and technology, the very things which have produced the environmental degradation humanity now faces. For example, Ian Bryce of the *Humanist Society of New South Wales*[81] made this speech of welcome to new members in 2014.

> The Humanist position includes the core idea that science provides the best path to reliable knowledge of our lives and the universe. This has not always been known - most ancient societies invented gods and religions. These helped to maintain the authority of the rulers, and purported to explain how the world worked. Many such belief systems came and went, reaching a peak in the Dark Ages. Sadly some linger on today.
>
> It was mainly in the Enlightenment period (starting around 1600) that the scientific method made its great advances ... These increasingly showed the religious claims to be myths. The churches of the day violently opposed these changes, imprisoning and executing many of the enlightened. Humanists reject claims of ancient gods, prophets and holy books. Refer to our meetings page for some views on the Enlightenment
>
> As well as mechanical things like the solar system, science has now addressed most aspects of life and humanity (and what used to be called spirituality or the supernatural), to the point where most of the Big Questions of philosophy have now been answered, or at least have been recast in modern terms, and indeed some new questions have arisen.[82]

---

81 Its Patrons include Dr Robyn Williams: "Robyn Williams AM (born 1944) is a science journalist and broadcaster resident in Australia who has hosted the Science Show on the Australian Broadcasting Corporation since 1975, Ockham's Razor (created 1984) and In Conversation (created 1997)".

82 http://www.hsnsw.asn.au/welcome.php

Against this exercise in blind optimism, others have been far less sanguine about the achievements of the scientific, industrial, and technological revolutions. Addressing the role of modern science in the ecological crisis, Vandana Shiva makes this observation.

> The sciences of maintaining life by maintaining the integrity of ecosystems have been rendered irrational by the reductionist mind. In their place have grown the sciences of disruption and destruction. The power of prediction was considered a unique power enjoyed by scientific knowledge. Increasingly, the predictions of science are falsified by the patterns unfolding in nature, because science is increasingly working across and against the patterns of nature. It is increasingly becoming a tool that destroys the balance between the manifold parts of nature which maintain the integrity of life on the planet ... The success of modern scientists has often been misleadingly linked to their offering solutions to more problems than ever before. However, it is rarely noticed that most of the problems modern science is occupied with are man-made and spin-offs of the transformative activity arising from prior beliefs.[83]

Secularist approaches to environmental sustainability

Professor Jeffrey D. Sachs is Special Advisor to the Secretary-General on the *Millennium Development Goals*, and Director of *The UN Millennium Project*. Professor Sachs presents the case for population control in what might misleadingly appear to be a moderate and appealing way. That case may be summed up in these terms:

1. World population growth is rising at such a rate that it will put unsustainable stresses on the resources of the planet.
2. Adding another 2.5 billion people to the planet will have catastrophic global consequences in terms of soaring rates of energy use which contribute to unsustainably greater levels of

---

83 Vandana Shiva, "Role of Modern Science in the Ecological Crisis", *Third World Resurgence*, no.16, Dec.1991, 2-5.

greenhouse gas emissions and to global warming.
3. Reducing population growth is an essential component (along with other measures) to meeting the challenge of global warming, mass starvation, "rapid deforestation, depletion of fisheries, land degradation, and the loss of habitat and extinction of a vast number of animal and plant species."[84]
4. Accordingly "population growth in developing regions – especially Africa, India, and other parts of Asia - needs to slow. Public policies can play an important role by extending access to family planning services to the poor, expanding social security systems, reducing child mortality through public health investments, and improving education and job opportunities for women."[85]
5. Dr Sachs concludes by setting out the advantages to human society and to the wider natural environment if population growth could be significantly slowed.

In his *UN Millennium Project Report to the UN Secretary-General*, the Director again emphasised the links between population, poverty and environmental degradation. But how is the slowing of population growth to be achieved? The answer being advocated by Dr Sachs and other population controllers is this: greater provision for "gender equality" measures which "include interventions for sexual and reproductive health, access to property rights and work, security, participation and institutional reform, and data collection and monitoring."[86]

Dr Sachs further specifies what he means by "gender equality" in these terms:

> *Universal access to sexual and reproductive health information and services and protection of reproductive rights.* ... Legislation and awareness campaigns to protect the rights of individuals and couples to plan their families; to ensure access to sexual

---

84 Jeffrey D Sachs, *The case for slowing population growth*, 2004, http://www.project-syndicate.org/commentary/the-case-for-slowing-population-growth
85 *Ibid.*
86 http://www.unmillenniumproject.org/documents/MainReportComplete-lowres.pdf

and reproductive health information and services; to discourage early marriage (at ages posing health risks), female genital mutilation, and other traditional harmful practices; and to expand access to safe abortions (where permitted by law) and review the legal status of abortion in order to improve public health while respecting national sovereignty, cultural values, and diversity.[87]

So the greater availability of contraception and "safe" abortion are key elements of gender equality, and "sexual and reproductive health".

The meaning of gender equality

There has been discussion and not a little confusion in the minds of many about just what the term "gender equality" really involves. It should be noted that the reference to diversity in the Sachs explanation of Gender Equality denotes the LGBTI agenda. As shown above, it is also clearly code for the widespread promotion and acceptance of contraception, "safe" abortion, and sterilisation.

In March 2010 the U.S. Secretary of State Hillary Clinton clarified the issue thus:

> If we're talking about maternal health, you cannot have maternal health without reproductive health. *And reproductive health includes contraception and family planning and access to legal, safe abortion.* [Emphasis added][88]

And "reproductive health" is also related to "gender equality" which in turn involves what Sachs has described as "universal access to sexual and reproductive health information and services and protection of reproductive rights."

---

87 *Ibid.*
88 http://www.brainyquote.com/quotes/quotes/h/hillarycli412151.html and https://www.youtube.com/watch?v=UH9rC0MaBJc

Sustainable Development

The United Nations convened a *Conference on Sustainable Development* held from the 20th to the 22nd of June 2012 in Rio de Janeiro, Brazil. "One of the main outcomes of the Rio+20 Conference was the agreement by member States to launch a process to develop a set of *Sustainable Development Goals* (SDGs), which will build upon the *Millennium Development Goals* and converge with the post 2015 development agenda. It was decided to establish an "inclusive and transparent intergovernmental process open to all stakeholders, with a view to developing global sustainable development goals to be agreed by the General Assembly".[89]

The goals of the *Sustainable Development* programme are similar to its immediate predecessor, the *Millennium Development Goals*. Among the 17 *Sustainable Development Goals* we find two references to the population control agenda in goals number three and number five.[90]

> **Goal 3:** Ensure healthy lives and promote well-being for all at all ages.[91]

The targets for goal 3 include the following:

> **3.7** by 2030 ensure universal access to sexual and reproductive health care services, including for family planning, information and education, and the integration of reproductive health into national strategies and programmes.[92]

> **Goal 5:** Achieve gender equality and empower all women and girls.[93]

---

89 https://sustainabledevelopment.un.org/topics/sustainabledevelopmentgoals

90 At the time of writing there are currently three draft sets of sustainable development goals however the Open Working Group (OWG) draft set of 17 goals is the most likely one to be developed when negotiations continue in June 2015.

91 https://sustainabledevelopment.un.org/focussdgs.html

92 ibid.

93 ibid.

Gender equality, as already pointed out involves "sexual and reproductive health information and services and protection of reproductive rights", that is abortion and contraception.

The targets for goal 5 include the following:

> **5.6** ensure universal access to sexual and reproductive health and reproductive rights as agreed in accordance with the Programme of Action of the ICPD and the Beijing Platform for Action and the outcome documents of their review conferences.

Engagement by the Catholic Church with the secular world

The Catholic Church has been in dialogue with worldly authorities from her beginning. Indeed Catholicism played a defining role in the development of Western culture. Her teaching on environmental issues is based on the book of Genesis written sometime in the fifteenth century BC (or perhaps as late as the 13th century BC[94]). Assuming Moses wrote Genesis (and that is only an assumption) the stories contained in the book could have been written between 1446 B.C. (the date of the Exodus) and 1406 BC (the death of Moses). Even then, some of the stories contained in Genesis could have originated much earlier, perhaps as far back as the second millennium BC.

By contrast, secular (non-religious) interest in environmental issues is of comparatively recent time.

> The origins of British environmentalism lie in the age of scientific discovery. The growth of interest in natural history revealed much about the consequences of man's exploitative relationship with nature. This led first to a movement to protect wildlife, and then to demands that rural amenity be provided as an antidote to life in the burgeoning industrial conurbations ... The first major

---

[94] Some scholars suggest it may have been written as late as the 8th or 9th centuries BC, but recognise that the stories would have been much older than that.

> influence on early British environmentalism was the study of natural history ... Gilbert White, whose seminal work *The Natural History of Selborne*, published in 1788, became the fourth most published book in the English language and influenced succeeding generations of naturalists, including Darwin.[95]

Of course the dialogue between Catholicism and science was of long standing. So it was natural that scientific discoveries would be incorporated into the Church's understanding of the environmentalism integral to the book of Genesis.

The alleviation of environmental degradation has come to be a central feature of Catholic social teaching on the environment in modern times, based upon her ancient theological tradition and enriched by the findings of modern science.

From the ancient theological tradition the Church sees the world as a great gift from God. Human beings, made in the image and likeness of God, are given stewardship of the rest of creation. Human beings are obligated by God to care for our planet. Central to this set of obligations is the prior obligation to protect the good of human life, its procreation and protection in the context of the family. Such obligations to the family are also to be found in the great secular documents of the United Nations, and especially in the *Universal Declaration on Human Rights* and its subsequent *Covenants*.[96] Here there is, on the face of it, agreement between Catholic social teaching and the secular world.

> Considering that, in accordance with the principles proclaimed in the Charter of the United Nations, recognition of the inherent dignity and of the equal and inalienable rights of all members of the human family is the foundation of freedom, justice and peace in the world,

---

95 John McCormick, *reclaiming Paradise: The Global Environmental Movement*, Indiana University Press, 1991, 2.

96 These foundational UN documents assert the inherent dignity of the human being from which emerge fundamental human rights. These rights include the right to life, the right to marry and found a family with the family being described as the natural and fundamental group unit of society.

Recognizing that these rights derive from the inherent dignity of the human person,[97]
1. The family is the natural and fundamental group unit of society and is entitled to protection by society and the State.
2. The right of men and women of marriageable age to marry and to found a family shall be recognized.[98]

Where the right to life is concerned, the UN documents apply to the unborn[99] and also to the rejection of euthanasia.[100]

It is important to note that, despite its commitment to fundamental human rights, various agencies of the United Nations continue to promote programmes which are fundamentally opposed to their own Charter and to Catholic moral teaching, programmes which cannot serve the common good. But the dialogue between the Church and the world must and will continue if the Church is to be faithful to her own Mission.

Further philosophical considerations

The Enlightenment project which forces "God" out of the drama of the lived experience of human beings delivers moral questions into the uncertain claims of competing moral philosophies and ultimately into moral subjectivism and relativism. While on the one hand there is a universal agreement on fundamental human values expressed as inviolable and inalienable rights, on the other

---

97 Preamble to the *International Covenant on Civil and Political Rights* [ICCPR]
98 ICCPR, Article 23.
99 Cf John I Fleming, "What Rights, if any, do the unborn have in international law?", in John Fleming and Nicholas Tonti-Filippini eds., *Common Ground?* St Pauls Publications, Strathfield, 2007, 266-291; John I Fleming and Michael Hains, "What Right , if any, do the unborn have in international law?" *Australian Bar Review* 16, n. 2 (November 1997) 181-98.
100 John I Fleming, "Euthanasia: Human Rights and Inalienability", *The Linacre Quarterly*, Volume 63, Number 1, February 1996, 44-56; 33. "I Cattolici Messi di Fronte Alle Strategie Pro-Eutanasia e Pro-Aborto", *Medicina E Morale*, 1996/1, 101-120; John I Fleming, "Death, Dying, and Euthanasia: Australia Versus The Northern Territory", *Issues in Law & Medicine*, 15:3, Spring 2000, 291-305.

hand these values are systematically undermined by those moral philosophies which purport to be scientific but which are, in fact, no more than a sleight of hand.[101]

Most moral philosophers typically proceed on the basis that human moral behaviour should conform to reason. The presumption is, of course, that reason itself can be mastered and controlled by the human intellect. But it is the failure of most moral philosophers to connect what they say with human practice which ensures a continuing estrangement between moral philosophy and the moral decisions people actually make as individuals and collectively.[102]

A good example of this is abortion. Philosophers who advocate abortion and infanticide place little to no emphasis on the mother's attachment to her child, an attachment which may be experienced as early as conception. To speak as if the attachment or bonding process begins only at birth or even a month or more later is unreal. The whole phenomenon of "childlessness" suggests a bonding to a child that is, as yet, no more than "a gleam in the parents' eyes".
To be "child-less" may mean that one does not have a child, but it also means that the couple does not have the child they already

---

101 Modern moral philosophy is in disarray, such that modern moral philosophers "offer a rhetoric which serves to conceal behind the masks of morality what are in fact the preferences of arbitrary will and desire." Cf Alasdair MacIntyre, *After Virtue*, London, Duckworth, 1987, 71.

102 Exchanges between Peter Singer and members of the Australian Senate Select Committee on the Human Embryo Experimentation Bill 1985 are instructive. In a discussion on whether embryos have intrinsic value, Singer asserted that an object or a person has intrinsic value if it is valued by someone else. Senator Crowley expressed surprise at this way of explaining intrinsic value. Singer observed that "it is always difficult, as a philosopher, to appreciate whether the way one is using terms is the way that they are used outside a discipline." Senator Crowley replied: "I thought it was fairly clear that philosophers do not use words like most everybody else in the community does or the community does not use them in the way philosophers do." *Senate Select Committee on the Human Embryo Experimentation Bill 1985, Official Hansard Report, 4 volumes,* Canberra: Australian Government Printing Service, 1986, 495. Later Senator Walters found herself in difficulties. "I do not know where to begin. I really did not know that philosophy was so far distant from community values, but I am learning." *Ibid.*, 501.

desire. The real lived experience of love, marriage and procreation out of which, no doubt, the universally held conviction of the inalienability and inviolability of human life has emerged, seems to have no place in much of what passes for post Enlightenment moral philosophy.[103] This may well explain why moral philosophies so often find themselves in opposition to, even hostile to, those universal convictions about basic human values.

By abstracting from the real human experience moral philosophers may find themselves at odds with the human conviction that there is, indeed, something special about human life as distinct from animal life. But if we can kill human beings in the name of, say, population control, then there is, in principle, no reason why the flesh meat of animals and humans may not be eaten if they die of natural causes. The idea of sending human bodies to be recycled as pet food offends the human conviction that the human body is to be treated with respect even after death, and properly cremated or buried according to local custom. If, however, the only morally relevant issues are sentience and certain other capacities experienced and in place, then as the animal liberation philosophy indicates, it is "speciesist" to imagine that there is something special about being human. If other dead animals can be recycled as pet food, then there is no reason, in principle, why dead human beings cannot be used in the same way.

The secular humanist trust in reason and science (the *sola fides* of scientific faith) promotes the abstraction of moral problems from their full human context. The solving of moral problems becomes a technical construct in which full credibility is given to the pretence that all relevant factors may be identified and in some way scientifically quantified.

Hence the popularity of utilitarianism where the good to be

---

[103] Exceptions to this include the natural law philosophy of John Finnis, Germain Grisez and Joseph Boyle. In his explication of the basic value of human life, Finnis includes "the transmission of life by procreation of children". John Finnis, *Natural Law and Natural Rights*, 86.

done or evil avoided is not simply identified with the act as an act, but is determined by a "scientific" calculation. That calculation requires computing the "weight" of pleasures achieved by the act over the "weight" of the pain caused. Of course such a "weighing" computation is nonsense. There are no scales by which we can add up pleasures and subtract pains. It is like trying to add a person's weight to their height, and subtracting the circumference of their midriff.

But utilitarianism is attractive to those minds which reflect these assumptions because it appears to be 'scientific', 'rational', and 'certain'. It appears to offer a scientifically safe method of computing morally good acts while at the same time appealing to the pragmatism of scientific elites who identify the public interest with the successful outcomes of their scientific protocols. Utilitarianism is able to simplify complex interactions brought about by human actions into the categories of pleasure and pain. What utilitarian philosophy does is to create a philosophical smoke-screen for inhuman behaviour by providing a philosophical and quasi-scientific justification for the elimination of human ballast, and for the imposition of a sectarian ethics on those who will be its victims.

As a back-up to these 'scientific' calculations in the event that they should fail to convince, utilitarians and, for that matter, many deontologists, with supreme confidence in reason, science and philosophy attempt to break the human agreement about fundamental human values by dividing the world into persons and non-persons. To be sure the elimination of persons may be justified by utilitarian calculations [*pace* human rights]. It is important to recognise this fact about utilitarianism, that the killing of innocent human beings may be justified in the interests of utility even if the ones to be killed are unarguably human persons. If, however, the expendable ones can be shown to be non-persons then we have a bioethics that is not the enemy of human rights, since, on this account, non-persons cannot be the bearers of human rights. In this way those moral philosophers who can justify killing human beings avoid the clash between their philosophies and the *consensus gentium* by defining some human individuals or classes of humans out of moral consideration.

But, *pace* Richard Dawkins *cum suis*, total trust in science and modern moral philosophy is misplaced. Such a trust is naïve as major philosophical figures from Cicero to Hobbes and to MacIntyre have noticed. In the seventeenth century Thomas Hobbes was skeptical about all previous moral philosophers and their philosophies. Citing the authority of Cicero, he asserted that "there can be nothing so absurd, but may be found in the books of philosophers".[104]

And in a devastating attack on the moral philosophy of his own time, which he described as "but a description of their own passions", Hobbes observed that:

> their logic, which should be the method of reasoning, is nothing else but captions of words, and inventions how to puzzle such as would go about to pose them. To conclude, there is nothing so absurd, that the old philosophers, as Cicero saith, (who was one of them,) have not some of them maintained.[105]

With much more humility Alasdair MacIntyre made a similar point about the moral philosophy of the twentieth century. MacIntyre argued that there: "is indeed a striking consensus against modern analytical philosophy concealed within it: every modern philosopher is against all modern moral philosophers except himself and his immediate allies. There is scarcely a need for an external attack."[106]

So, when philosophers attempt to construct a total view based on their own philosophical assumptions, and purposely excluding the insights of theology, there is bound to be difficulty and tragedy.

---

[104] Thomas Hobbes, *Leviathan, or the Matter, Forme and Power of a Commonwealth Ecclesiastical and Civil*, ed. M. Oakeshott, Oxford: Blackwell, 1960, 27.

[105] *Ibid.*, 438-439. Hobbes completes the attack by debunking Aristotelianism. Aristotle's Metaphysics he calls "absurd", his Politics "repugnant to government", and a "great part of his Ethics" ignorant.

[106] Alasdair MacIntyre, "Why Is the Search for the Foundation of Ethics So Frustrating?", *Hastings Center Report*, 9:4, August 1979, 18.

Saint John Paul II put it this way:

> In particular, it is necessary to keep in mind the unity of truth, even if its formulations are shaped by history and produced by human reason wounded and weakened by sin. *This is why no historical form of philosophy can legitimately claim to embrace the totality of truth, nor to be the complete explanation of the human being, of the world and of the human being's relationship with God.*[107] [Emphasis added]

On the other hand, Saint John Paul II argued for that special relationship between faith and reason which is productive of true wisdom.

> Reflecting in the light of reason and in keeping with its rules, and guided always by the deeper understanding given them by the word of God, Christian philosophers can develop a reflection which will be both comprehensible and appealing to those who do not yet grasp the full truth which divine Revelation declares. *Such a ground for understanding and dialogue is all the more vital nowadays, since the most pressing issues facing humanity – ecology, peace and the co-existence of different races and cultures, for instance – may possibly find a solution if there is a clear and honest collaboration between Christians and the followers of other religions and all those who, while not sharing a religious belief, have at heart the renewal of humanity.* The Second Vatican Council said as much: "For our part, the desire for such dialogue, undertaken solely out of love for the truth and with all due prudence, excludes no one, neither those who cultivate the values of the human spirit while not yet acknowledging their Source, nor those who are hostile to the Church and persecute her in various ways". A philosophy in which there shines even a glimmer of the truth of Christ, the one definitive answer to humanity's problems, will provide a potent underpinning for the true and planetary ethics which the world now needs.[108] [Emphasis added]

---

107 Encyclical Letter *Fides et ratio* (1998), of Pope John Paul II to the Bishops of the Catholic Church on the Relationship Between Faith and Reason, n. 51.
108 *Ibid.*, n. 104.

This positive and realistic approach to the good of philosophy while recognising its limits, provides the entrée for the contribution to public debate on current environmental issues by previous popes and lately by Pope Francis in his *Laudato Si'*.

CHAPTER 3

What is Catholic social teaching?

From its beginning, the early Church had to spell out how Christians could live in the world and be faithful to the Gospel. As the resurrected Lord Jesus Christ was about to ascend into heaven, he gave his eleven remaining disciples their final instructions. He knew some of them still had their doubts. Even so, he still spelt out their task:

> All authority in heaven and on earth has been given to me. Go therefore, and make disciples of all nations, baptising them in the name of the Father, and of the Son, and of the Holy Spirit., teaching them to observe all that I have commanded you; and lo, I am with you always, to the close of the age.[109]

At first the eleven hid themselves away from the world. Apart from their doubts, they were afraid of what would happen to them once they came to the attention of the authorities. Moreover, they knew they had to wait to "be baptised with the Holy Spirit".[110] And soon, on the day of Pentecost they received the Holy Spirit, the Church was born, and the Apostles, transformed by the Holy Spirit, immediately and courageously set about doing what they had been told to do.

It was not long before the Apostles had to face up to the reality of the world and how a Christian was to live in a world hostile to the Gospel. They had been prepared for this. The Apostles remembered that they had been given authority to safeguard the teachings of Jesus to interpret the Gospel, and make definitive teaching. Apart from the theological and doctrinal issues which were to be dealt

---
109 Matthew 28: 18-20.
110 Acts 1:5.

with as well, they had to spell out the moral consequences of embracing the Gospel.

It is crucial to remember that the starting point for this reflection was the faith itself, not the agenda of the world. This was not about being relevant to the world, as some would have us act in these days. The world does not set the agenda for the Church. The faith is the starting-point for this reflection, not simply concern about particular issues facing society.

Nevertheless, the facts of life had to be dealt with, and the early Church had to find a way to live in the context of Judaism. They set down what was to be the religious practices which would govern their daily lives. So they embraced these activities:

1. They devoted themselves to learning the Gospel at the feet of the Apostles;
2. They kept up a daily attendance at the Temple;
3. In their homes the Eucharist was celebrate daily;
4. They often ate meals together to cement the life of the Christian fellowship; and[111]
5. Their attitude to their worldly goods was completely transformed. They had to discover what it meant to belong to the community of their fellow Believers so they shared their worldly goods with the Christian community.[112]

Moreover, they had to deal with the reality of the kinds of food available in the markets and individual personal preference. One person will eat virtually anything while another is a vegetarian. St Paul taught them that they were to learn not to fight over such matters.[113]

But other dietary matters were more problematic. While non-Jewish men who had become Christians were not required to undergo circumcision, nevertheless they were to abstain from food deriving from idol worship.

---

111 Acts 2:42-47.
112 Acts 4:32-37.
113 Romans 14:1.

To deal with the problem of "emperor worship", St Paul urges his fellow believers only to pray for all those in high places including kings, and to live peaceably and quietly so as to be a good example to others.[114] And in a society which practiced slavery, they were to behave differently. All Christians, whether slave or free, were equal before God.[115] And St Paul writes to the Christian owner of the runaway slave, Onesimus, prevailing on him to receive his slave back without punishment, and not as a slave but as a fellow Christian.[116]

Over the course of centuries, as the Church became stronger, she managed through her teaching to overturn the practice of slavery.[117] In the 11th, 12th, and 13th centuries the Catholic Church developed its "Just War" theory precisely to restrict the ambitions and actions of various European countries.

In chapter 1 I have already referred to the gradual evolution of the Encyclical as a mode of communication used by popes. But a major development, perhaps *the* major development in Catholic social teaching, occurred on the 15th of May 1891 when Pope Leo XIII released his Encyclical letter *Rerum novarum* on the "Rights and Duties of Capital and Labour". Written at the height of the industrial revolution and the challenge of Marxist socialism, and in the context of manifest injustice where the working classes were concerned, the Pope wrote an open letter, passed to all Catholic bishops, which addressed the condition of the working classes. It was a clarion call for recognition of injustice and how the Church should respond.

> That the spirit of revolutionary change, which has long been disturbing the nations of the world, should have passed beyond the sphere of politics and made its influence felt in the cognate

---

114 1 Timothy 2:1-4.
115 Galatians 3:28.
116 Philemon 8-20.
117 *Cf* Pope Eugene IV, *Sicut dudum*, 1435; Pope Paul III: *Sublimis Deus*, 1537; Pope Gregory XVI: *In Supremo*, 1839.

sphere of practical economics is not surprising. The elements of the conflict now raging are unmistakable, in the vast expansion of industrial pursuits and the marvellous discoveries of science; in the changed relations between masters and workmen; in the enormous fortunes of some few individuals, and the utter poverty of the masses; the increased self-reliance and closer mutual combination of the working classes; as also, finally, in the prevailing moral degeneracy. The momentous gravity of the state of things now obtaining fills every mind with painful apprehension; wise men are discussing it; practical men are proposing schemes; popular meetings, legislatures, and rulers of nations are all busied with it – actually there is no question which has taken deeper hold on the public mind.[118]

Hence, by degrees it has come to pass that working men have been surrendered, isolated and helpless, to the hardheartedness of employers and the greed of unchecked competition. The mischief has been increased by rapacious usury, which, although more than once condemned by the Church, is nevertheless, under a different guise, but with like injustice, still practiced by covetous and grasping men. To this must be added that the hiring of labour and the conduct of trade are concentrated in the hands of comparatively few; so that a small number of very rich men have been able to lay upon the teeming masses of the labouring poor a yoke little better than that of slavery itself.[119]

To remedy these wrongs the socialists, working on the poor man's envy of the rich, are striving to do away with private property, and contend that individual possessions should become the common property of all, to be administered by the State or by municipal bodies. They hold that by thus transferring property from private individuals to the community, the present mischievous state of things will be set to rights, inasmuch as each citizen will then get his fair share of whatever there is to enjoy. But their contentions are so clearly powerless to end the controversy that were they carried into effect the working man

---

118 Pope Leo XIII, *Rerum novarum*, n. 1.
119 *Rerum novarum*, n. 3.

> himself would be among the first to suffer. They are, moreover, emphatically unjust, for they would rob the lawful possessor, distort the functions of the State, and create utter confusion in the community.[120]

Such was the manifest influence of this crucial development in Catholic social teaching that Saint John Paul II, marking the centenary of *Rerum novarum*, would describe the centenary as "an occasion of great importance for the present history of the Church and for my own Pontificate".[121] He spoke of "the debt of gratitude which the whole Church owes to this great Pope and his 'immortal document'".[122]

> Although the commemoration at hand is meant to honour *Rerum novarum*, it also honours those Encyclicals and other documents of my Predecessors which have helped to make Pope Leo's Encyclical present and alive in history, thus constituting what would come to be called the Church's "social doctrine", "social teaching" or even "social magisterium".[123]

The ongoing importance of the social doctrine taught by his predecessors was reiterated by Saint John Paul II, noting that:

> The validity of this teaching has already been pointed out in two Encyclicals published during my Pontificate: *Laborem exercens* on human work, and *Sollicitudo rei socialis* on current problems regarding the development of individuals and peoples.[124]

---

120 *Reum novarum*, n. 4.
121 Pope John Paul II, *Centesimus annus*, 1st May 1991, n. 1.
122 *Centesimus annus*, n. 1. The expression "immortal document" comes from Pope Pius XI, *Quadragesimo anno*, 15th May 1931, n. 39.
123 *Centesimus annus*, n. 2.
124 *ibid.*

What then is Catholic social teaching?

Saint John Paul II defined Catholic social teaching in this way:

> The Church's social doctrine is not a "third way" between liberal capitalism and Marxist collectivism, nor even a possible alternative to other solutions less radically opposed to one another: rather, it constitutes a category of its own. *Nor is it an ideology, but rather the accurate formulation of the results of a careful reflection on the complex realities of human existence, in society and in the international order, in the light of faith and of the Church's tradition.* Its main aim is to interpret these realities, determining their conformity with or divergence from the lines of the Gospel teaching on man and his vocation, a vocation which is at once earthly and transcendent; its aim is thus to guide Christian behaviour. It therefore belongs to the field, not of ideology, but of theology and particularly of moral theology.[125] [Emphasis added]

So, while Catholic social teaching is not a political or economic ideology or programme, it is in a special category of its own:

- It is an exercise in moral theology;
- It is a careful reflection on what it means to be a human being in the contemporary era which it addresses and in the light of the Church's tradition about the nature of man, and in the light of faith;
- Its aim is to provide a guide as to the way in which we should live in the light of man's vocation as both a citizen of this world and his vocation to be a citizen of the next.

Indeed, as William Newton points out, Catholic social teaching is founded upon an adequate understanding of the nature of the human being, that man is made in the image of God, that he is a fallen creature prone to sin, and that Christ has overcome the power of sin enabling us to have a reasonable hope that man can change the way he has so far constructed his social life. Based upon that adequate anthropology, Catholic social teaching has three other principles:

---

125 *Sollicitudo rei socialis*, n. 41.

These are the common good, subsidiarity, and solidarity.[126]

These principles are all interrelated and are "applicable to all spheres of social doctrine, whether economics, politics, international community, or the environment".[127] For a fuller explication of these principles William Newton's book here referenced is an excellent resource.

But briefly the **common good** refers to those goods necessary for full flourishing of all human beings. The *Catechism of the Catholic Church* speaks of the "common good" in these terms:

> The common good comprises "the sum total of social conditions which allow people, either as groups or as individuals, to reach their fulfillment more fully and more easily" (*Gaudium et spes*, n. 26).
>
> The common good consists of three essential elements: respect for and promotion of the fundamental rights of the person; prosperity, or the development of the spiritual and temporal goods of society; the peace and security of the group and of its members.
>
> The dignity of the human person requires the pursuit of the common good. Everyone should be concerned to create and support institutions that improve the conditions of human life.
>
> It is the role of the state to defend and promote the common good of civil society. The common good of the whole human family calls for an organization of society on the international level.[128]

Subsidiarity is defined in the *Catechism of the Catholic Church*:

> Certain societies, such as the family and the state, correspond more directly to the nature of man; they are necessary to him. To promote the participation of the greatest number in the life of a society, the creation of voluntary associations and institutions must be encouraged "on both national and international levels,

---

126 William Newton, *A Civilization of Love*, Leominster, UK, 2011, 26.
127 *Ibid.*
128 *Catechism of the Catholic Church*, nn 1924-1927.

which relate to economic and social goals, to cultural and recreational activities, to sport, to various professions, and to political affairs." This "socialization" also expresses the natural tendency for human beings to associate with one another for the sake of attaining objectives that exceed individual capacities. It develops the qualities of the person, especially the sense of initiative and responsibility, and helps guarantee his rights.[129]

Socialization also presents dangers. Excessive intervention by the state can threaten personal freedom and initiative. The teaching of the Church has elaborated the principle of subsidiarity, according to which "a community of a higher order should not interfere in the internal life of a community of a lower order, depriving the latter of its functions, but rather should support it in case of need and help to co- ordinate its activity with the activities of the rest of society, always with a view to the common good."[130]

Solidarity is defined in the *Catechism of the Catholic Church* in these terms:

> The principle of solidarity, also articulated in terms of "friendship" or "social charity," is a direct demand of human and Christian brotherhood.
>
> An error, "today abundantly widespread, is disregard for the law of human solidarity and charity, dictated and imposed both by our common origin and by the equality in rational nature of all men, whatever nation they belong to. This law is sealed by the sacrifice of redemption offered by Jesus Christ on the altar of the Cross to his heavenly Father, on behalf of sinful humanity."[131]
>
> Solidarity is manifested in the first place by the distribution of goods and remuneration for work. It also presupposes the effort for a more just social order where tensions are better able to be reduced and conflicts more readily settled by negotiation.[132]
>
> Socio-economic problems can be resolved only with

---

129 *Catechism of the Catholic Church*, n. 1882.
130 *Ibid.*
131 *Ibid.*, n. 1939.
132 *Ibid.*, n. 1940.

the help of all the forms of solidarity: solidarity of the poor among themselves, between rich and poor, of workers among themselves, between employers and employees in a business, solidarity among nations and peoples. International solidarity is a requirement of the moral order; world peace depends in part upon this.[133]

The virtue of solidarity goes beyond material goods. In spreading the spiritual goods of the faith, the Church has promoted, and often opened new paths for, the development of temporal goods as well. And so throughout the centuries has the Lord's saying been verified: "Seek first his kingdom and his righteousness, and all these things shall be yours as well".

For two thousand years this sentiment has lived and endured in the soul of the Church, impelling souls then and now to the heroic charity of monastic farmers, liberators of slaves, healers of the sick, and messengers of faith, civilization, and science to all generations and all peoples for the sake of creating the social conditions capable of offering to everyone possible a life worthy of man and of a Christian.[134]

The Church's authority where fundamental moral questions are concerned, whether or not infallibly proposed, is binding on all Catholics. Social teaching involves the application of these moral principles to the realities of the world which the particular magisterial document addresses. With regard to papal documents, the content and context of the document is also important, as is the way that the Pope expresses himself. We should therefore ask ourselves the following questions:

- Is the statement solemnly expressed?
- Is it repeating something taught consistently by the Church from the earliest times?
- Is it repeating something of natural law?
- Is the intention to define, or merely to propose and exhort?[135]

---

133 *Ibid.*, n. 1941.
134 *Ibid.*, n. 1942.
135 William Newton, *A Civilization of Love*, Leominster, UK, 2011, 8.

In the case of *Laudato Si'*, the Pope describes his document as a "reflection" and a "proposal".

> *In light of this reflection, I will advance some broader proposals for dialogue and action* which would involve each of us as individuals, and also affect international policy. Finally, convinced as I am that change is impossible without motivation and a process of education, I will offer some inspired guidelines for human development to be found in the treasure of Christian spiritual experience.[136] [Emphasis added]

Laudato Si' is in no sense, then, to be regarded as a political programme, an ideology, an infallible mandate for particular political and social action as I will discuss in chapter 6 below in relation to certain Church agencies.

Another formulation of the basis of Catholic social teaching

Another way of describing the foundational principles of Catholic Social Teaching (CST) has been admirably set out by Dr Michael Casey and Father Peter Smith from the Archdiocese of Sydney. The advantage of this formulation is the clearer emphasis on the "dignity of the human person" which leads to *the preferential option for human life*. The principles of solidarity and the common good lead to the Church's preferential option for the poor.

1. **The Dignity of the Human Person**: The focal point of CST is the human person, made in the image and likeness of God, and so having fundamental freedom and dignity which in turn is the basis for human rights. Recognising this image in our neighbour, CST rejects any policy that reduces people to economic units or passive dependence.
2. **The Common Good**: The common good refers to what is good for all people. We exist as part of society. Every individual has a duty to share

---

136 *Laudato Si'*, n. 15.

in promoting the welfare of the community and the right to benefit from that welfare. This applies at every level: local, national, and international. Public authorities exist mainly to promote the common good and to ensure that no section of the population is excluded.

3. **Solidarity**: Solidarity is standing with others. As members of the one human family we have mutual obligations to promote the rights and development of peoples across communities and nations. Solidarity is a fundamental bond of unity and interdependence between people. All are responsible for all, and in particular the rich have responsibilities towards the poor. National and international structures must reflect this.

4. **Subsidiarity**: Subsidiarity means that all power and decision-making in society should be at the most local level compatible with the common good. This will often mean decision-making and power passing downwards, but it can also mean passing appropriate authority upwards. The balance between the vertical dimension of social justice (subsidiarity) and the horizontal dimension (solidarity) is achieved by keeping the common good clearly in focus.

5. **Option for the Poor**: The option for the poor means that we should put the poor at the forefront of decisions we make. This was implicit in the early development of CST and has now been taken up with new urgency and far-reaching consequences for pastoral action. For Christians, fidelity to Christ means seeing him above all in the faces of suffering and wounded people and the poor.[137,138]

To arrive at these principles the Church has had regard to the law of the Gospel and also to the natural moral law. Or as Cardinal Angelo Sodano put it, "The principles of the Church's social doctrine, which are based on the natural law, are then seen to be confirmed and strengthened, in the faith of the Church, by the Gospel of

---

137 *Compendium of the Social Doctrine of the Church*, Libreria Editrice Vaticana, 2004, 91-112.
138 M Casey and P Smith, "Catholic Social Teaching and Papal Encyclicals", op.cit.

Christ."[139]

Thus, not only with respect to the Law of the Gospel but also where the natural law is concerned, the Catholic Church has the authority to correctly interpret the moral law and its application to concrete social situations. In his Encyclical Letter, *Humanae vitae*, and in the section labelled "Competency of the Magisterium", Blessed Paul VI makes this position clear:

> This kind of question requires from the teaching authority of the Church a new and deeper reflection on the principles of the moral teaching on marriage – a teaching which is based on the natural law as illuminated and enriched by divine Revelation.
>
> No member of the faithful could possibly deny that the Church is competent in her magisterium to interpret the natural moral law. It is in fact indisputable, as Our predecessors have many times declared, (1) that Jesus Christ, when He communicated His divine power to Peter and the other Apostles and sent them to teach all nations His commandments, (2) constituted them as the authentic guardians and interpreters of the whole moral law, not only, that is, of the law of the Gospel but also of the natural law. For the natural law, too, declares the will of God, and its faithful observance is necessary for men's eternal salvation.

The limits of Catholic social teaching

Of course there are limits to the teaching authority of the Pope. These limits were clearly explained, in 1998, by the Congregation for the Doctrine of the Faith, in response to a request by Saint John Paul II.

> The Roman Pontiff – like all the faithful – is subject to the Word of God, to the Catholic faith, and is the guarantor of the Church's obedience; in this sense he is *servus servorum Dei*. He does not make arbitrary decisions, but is spokesman for the will of the Lord, who speaks to man in the Scriptures lived and interpreted

---

[139] Cardinal Angelo Sodano, *Compendium of the Social Doctrine of the Church*, Libreria Editrice Vaticana, 2004, xxiii.

by Tradition; in other words, the *episkope* of the primacy has limits set by divine law and by the Church's divine, inviolable constitution found in Revelation. The Successor of Peter is the rock which guarantees a rigorous fidelity to the Word of God against arbitrariness and conformism: hence the martyrological nature of his primacy.[140]

Principles and Values

Catholic social teaching not only proceeds on the basis of the principles which guide its methodology, it is also informed by fundamental human values, values which are inherent in the dignity of the human person. This approach is also clear in the United Nations *International Covenant on Civil and Political Rights* which, in its *Preamble*, says this:

> Considering that, in accordance with the principles proclaimed in the Charter of the United Nations, recognition of the inherent dignity and of the equal and inalienable rights of all members of the human family is the foundation of freedom, justice and peace in the world.
>
> Recognizing that these rights derive from the inherent dignity of the human person ...[141]

These fundamental values refer to the inviolable and inalienable right of every human being to life, freedom, truth and justice. And here justice refers not just to justice for the individual but in the way in which classes of people are treated in accordance with the common good of all.

---

140 Congregation for the Doctrine of the Faith, *The Primacy of the Successor of Peter in the Mystery of the Church*, n. 7. *L'Osservatore Romano*, Weekly Edition in English, 18 November 1998, 5-6.
141 *International Covenant on Civil and Political Rights*, United Nations 1966, *Preamble*.

The Church's right to speak[142]

Some people object to the Church entering into the discussion on issues such as environmental ethics, abortion, marriage, divorce, reproductive technology, and the rights of workers. This objection is due to the failure either to know about or to understand the basis upon which the Church makes her contribution to the common good of all.

Simply expressed the objection runs something like this.

- Religion is a matter of private opinion.
- The Church's moral opinions are based upon a religion which itself is purely a private affair between the believer and that religion.
- Christianity is not objectively true, but purely a subjective choice based upon a number of non-rational considerations, that is to say, considerations not themselves subject to rational scrutiny.
- Moreover, the Church should not be able to force its opinions onto the community. In fact it would be better if the Church was neither seen nor heard in the public square where the real debates on morality and the law take place.

In the United States of America the doctrine of separation of Church and State has been turned on its head. The founders of the US had in mind the protection of the Church from the State and not vice versa. But today, in the US, their Constitution has been interpreted to restrict the rights of religious believers to act publicly in accordance with their religious convictions. Freedom *of* religion has become freedom *from* religion. In Australia, the historical development of the relationship between Church and State has been different. This development is linked to the Australian Constitution. Section 116 provides:

> The Commonwealth shall not make any law for establishing any religion, or for imposing any religious observance, or for prohibiting the free exercise of any religion, and no religious test

---

142 Much of this section has been taken directly from John Fleming and Nicholas Tonti-Filippini, "Seeking a consensus", *Common Ground?*, St Pauls Publications, Strathfield, NSW, 2007, 322-325.

shall be required as a qualification for any office or public trust under the Commonwealth.

The meaning of section 116 was determined by the High Court of Australia in the famous "Defence of Government Schools" (DOGS) case in 1981. Barwick CJ: "the establishment of religion must be found to be the object of the making of the law. Further, because the whole expression is "for establishing any religion", the law to satisfy the description must have that objective as its express and, as I think, single purpose.[143]

The purpose of these provisions in the Australian Constitution is, then, to limit the role of the State, not the Church or any other religious grouping. Having come from a society where the King nationalised religion and made the Church a department of State under parliamentary control, persecuting and marginalising those whose religious opinions differed from those of the State, it is not surprising that the founders wanted a Constitution which would allow maximum freedom of religion. Where religion is concerned it is the Church that needs protection from the hubris of politicians and not vice versa. The Church did not impose religion upon England. England imposed its views on the Church.

Moreover, the Australian Constitution does not exclude religious arguments, nor does it exclude religious people or the Churches from public debate. The opposite is true. People are not to have their religious freedom infringed by the state and are not to be inhibited from expressing their religious opinions in the public square. The *Australian Constitution* itself recognises the legitimacy of religion in the public square when, in its Preamble, it says that we, the Australian people, are "humbly relying on the blessings of Almighty God". This is further supported by the custom of the Parliament to begin each day with prayer including the "Our Father".

Perhaps it is fairer to say that the Australian Constitution provides for the cooperation between church and state, – or, better in our multicultural community – religion and state. Michael Hogan,

---

[143] http://www.austlii.edu.au/au/cases/cth/HCA/1981/2.html.

Research Associate in Government and International Relations at the University of Sydney, put it this way:

> Australia does not have a legally entrenched principle, or even a vague set of conventions, of the separation of church and state. From the appointment of Rev. Samuel Marsden as one of the first magistrates in colonial New South Wales, to the adoption of explicit policies of state aid for denominational schools during the 1960s, to the two examples mentioned above, Australia has had a very consistent tradition of cooperation between church and state. 'Separation of church and state', along with 'the separation of powers' or 'pleading the Fifth', are phrases that we have learned from the US, and which merely serve to confuse once they are taken out of the context of the American Constitution.
>
> What Australia does have is a principle of state *neutrality*, or equal treatment, when dealing with churches. This principle dates back at least to Governor Bourke (if not to Macquarie) in colonial NSW, and extends all the way into contemporary Australia where government monies at all levels go quite happily to the churches so that they can run schools, hospitals, employment agencies, social welfare bureaux and even drug injecting rooms. This principle of neutrality is not entrenched in either the State or Federal Constitutions, and has no legal standing. (Constitutionally, State governments could still conceivably nominate an established church; only the Commonwealth is forbidden to do so by Section 116 of its Constitution!) Ultimately, the strength of the principle comes from the conventions hammered out in colonial Australia that saw English and Scottish established churches deprived of their priority in government funding. It survives into the twenty-first century because no major party could seriously contemplate abandoning it.
>
> The principle of state neutrality has coexisted in Australia with a strong secular tradition in politics. ...For most of our history most Australians have been quite happy with the principle

that governments should not favour one church over another.[144]

Notwithstanding the legal position, many politicians and others have behaved in a way that does not respect the *Australian Constitution* by demanding that bishops, priests, ministers, churches, and other religious bodies stop "meddling" in politics. Such *ad hominem* attacks represent an egregious appeal to prejudice and unjust discrimination against certain people or institutions. It is also hypocritical in the strict sense because such advice is usually given by, but not expected to apply to, those whose religion is variously described as secular, 'humanist', atheistic or agnostic.

In at least one Australian State, though, the right for Catholics to teach both by word and example are under threat at the time of writing. The *'bien pensants'*, who claim to know what is really best for Australia, use legal means to close down public debate on contentious social-moral questions such that the teaching authority of the Catholic Church has to be silenced.

In Tasmania the Catholic Archbishop of Hobart, the Most Reverend Julian Porteous, has had action taken against him for allegedly breaching section 17(1) of *The Tasmanian Anti-Discrimination Act* 1998. The complainant is Martine Delaney, a man who "changed" into a woman. Delaney is a Greens candidate for the Federal seat of Franklin. The substance of her complaint concerns a pastoral letter by the Archbishop on the subject of same-sex marriage. Delaney complains that (1) she was offended by the letter, (2) that the offending part of the letter referred to same-sex relationships as a "friendship" and that same sex partners were not spouses thereby saying that same-sex partners did not deserve equal recognition, and (3) that a reasonable person would have anticipated

---

144 Michael Hogan, Separation of church and state?, 16 May 2001, http://www.australianreview.net/digest/2001/05/hogan.html

that Delaney would have been offended by the pastoral letter.[145]

In Australia the government has decided to put the matter of same sex marriage to a vote of the people in a plebiscite. But many proponents of same-sex marriage do not want the matter put to the people in a plebiscite because "we're just going to give a voice to the worst elements of our society, the bigots and bullies, putting them on our TVs, putting them on our radio and subjecting some of the most vulnerable people in our society to those hateful opinions."[146]

Bill Shorten, the Leader of the Opposition in the Australian Parliament believes that an issue like same sex marriage can only be debated and voted on by the political class. Australians in general, simply cannot be trusted to debate issues such as these.

> A plebiscite could act as a lightning rod for the very worst of the prejudice so many LGBTI Australians endure. A platform for people to attack, abuse and demean Australians on the basis of who they love. The fact is, casual, unthinking discrimination and deliberate, malicious homophobia are still far too common in our society. It's not confined to keyboard warriors and Twitter trolls. It's in our schoolyards, our workplaces, our sporting clubs. This takes a heavy toll on mental health, particularly for young people.[147]

Apparently it is only politicians who have the sophistication and *gravitas* to properly debate and vote on redefining marriage and family, even though it is families who constitute the state, not the

---

145 Robin Speed, "Appeal to freedoms will not avail Archbishop", *News Weekly*, 24th October 2015, 5-6. Robin Speed is president of the Rule of Law Institute Australia. Speed says that the interest of the Rule of Law Institute is solely in the legal aspects of this case and "makes no comment on the merits of changing the definition of marriage."
146 Australian Labor Party candidate, Pat O'Neill, http://www.smh.com.au/federal-politics/political-news/samesex-plebiscite-a-beacon-for-bullies-and-bigots-says-gay-candidate-20151022-gkg9mo.html
147 Bill Shorten, "Gay marriage: Malcolm Turnbull's $140m plebiscite risks a platform for abuse", opinion piece published in the *Sydney Morning Herald*, 22 October 2015, http://www.smh.com.au/federal-politics/political-opinion/gay-marriage-malcolm-turnbulls-140m-plebiscite-risks-a-platform-for-abuse-20151022-gkfg5r.html

state that defines families.

But politicians and media commentators often display bigotry and hubris in dealing with social issues. Examples of publicly expressed religious bigotry by significant members of the press, political establishment, and others abound. The views of Christians are associated with fundamentalism, that unenlightened and ignorantly dogmatic religion, which is impervious to science, reason, and compassion.

Alex Mitchell, columnist for Sydney's *The Sun Herald*, exemplified the crudest expression of anti-Catholic bigotry when accounting for the way in which NSW Senators voted against a private member's bill to overturn the ban on therapeutic cloning in 2006. Senators Ursula Stephens and Steve Hutchins were described as coming "from the darkest recesses of the NSW right", while Senators Bill Heffernan and Concetta Fierravanti-Wells were "mediaevalists" who "took their stand somewhere around the 15th century when the Spanish Inquisition was in full swing".[148]

Senator Amanda Vanstone, in supporting therapeutic cloning, said: "There are different views on when life begins, but no religion has the right to seek to have its view legislated." Never mind that Senator Vanstone then voted to have her own personal secular religious views legislated. Each politician had to vote, and Vanstone cast her vote according to her own opinion. But she was wrong to tell politicians of a different religious opinion from her own that they did not have the same right to seek to persuade the parliament to that other point of view. There is nothing in the *Australian Constitution* to justify the denial of equal rights to free speech on the basis of a person's religious or other practices and opinions.

It is also worth noting that when the Church says something with which media commentators and politicians agree, then that is alright for the Church to speak her mind. Where *Laudato Si'* is concerned,

---

[148] Alex Mitchell, "Faulkner lone state ALP senator to back cloning legislation", *The Sun Herald*, 12 November 2006, 22.

commentators who agree with the Pope laud his encyclical. He is even to be congratulated even if in a somewhat patronising way! On the other hand, those who oppose what the Pope says suggest it would be better if the Pope stayed out of matters that don't concern him. For example, in the context of US politics, when the Pope declared the Vatican's support for a separate Palestinian state, saying that it would recognise such a state, Rep. Jeff Duncan (R-S.C.), a hawkish defender of Israel, said: "It's interesting how the Vatican has gotten so political when ultimately the Vatican ought to be working to lead people to Jesus Christ and salvation, and that's what the Church is supposed to do".[149] In other words, Holy Father, stick to the spiritual realms and leave the politics to us. But, as Steve Benen says:

> When church leaders condemn abortion, congressional Republicans shout, "Amen". When the Pope enters a foreign policy debate, suddenly we effectively hear, "Mind your own business, padre."[150]

Against this background of intolerance, bigotry, and attempts to silence the Church on social issues, Pope Francis continues the Catholic tradition of reiterating the fundamental principles and moral values that should underpin public policy. That tradition has been explained by the United States Conference of Catholic Bishops in these terms:

> Catholic social teaching is a central and essential element of our faith. Its roots are in the Hebrew prophets who announced God's special love for the poor and called God's people to a covenant of love and justice. It is a teaching founded on the life and words of Jesus Christ, who came "to bring glad tidings to the poor ... liberty to captives ... recovery of sight to the blind"(Lk 4:18-19), and who identified himself with "the least of these," the hungry and the stranger (cf. Mt 25:45). Catholic social teaching is built

---

149 Steve Benen, "House Republican: Pope should stay out of politics", http://www.msnbc.com/rachel-maddow-show/house-republican-pope-should-stay-out-politics
150 *Ibid.*

on a commitment to the poor. This commitment arises from our experiences of Christ in the Eucharist.

As the Catechism of the Catholic Church explains, "To receive in truth the Body and Blood of Christ given up for us, we must recognize Christ in the poorest, his brethren" (no. 1397).

Catholic social teaching emerges from the truth of what God has revealed to us about himself. We believe in the triune God whose very nature is communal and social. God the Father sends his only Son Jesus Christ and shares the Holy Spirit as his gift of love. God reveals himself to us as one who is not alone, but rather as one who is relational, one who is Trinity. Therefore, we who are made in God's image share this communal, social nature. We are called to reach out and to build relationships of love and justice.

Catholic social teaching is based on and inseparable from our understanding of human life and human dignity. Every human being is created in the image of God and redeemed by Jesus Christ, and therefore is invaluable and worthy of respect as a member of the human family. Every person, from the moment of conception to natural death, has inherent dignity and a right to life consistent with that dignity. Human dignity comes from God, not from any human quality or accomplishment.

Our commitment to the Catholic social mission must be rooted in and strengthened by our spiritual lives. In our relationship with God we experience the conversion of heart that is necessary to truly love one another as God has loved us.[151]

To ask the Church to be silent in the face of the serious moral questions of the day is akin to asking Pope Francis (or any pope) to be like the priest and Levite in the story of the Good Samaritan.[152] Just ignore the serious social issues and avert your holy eyes from damaged people even when they are right under your nose for fear that the malefactors may attack you!

A full and yet concise expression of the Church's teaching

---

151 http://www.usccb.org/beliefs-and-teachings/what-we-believe/catholic-social-teaching/sharing-catholic-social-teaching-challenges-and-directions.cfm
152 Luke 10:25-37.

on social justice may be found in the *Catechism of the Catholic Church*, at Article 3. At the heart of the Church's impulse to share what it has to the world is this:

> Social justice can be obtained only in respecting the transcendent dignity of man. The person represents the ultimate end of society, which is ordered to him:
> What is at stake is the dignity of the human person, whose defense and promotion have been entrusted to us by the Creator, and to whom the men and women at every moment of history are strictly and responsibly in debt.
> Respect for the human person entails respect for the rights that flow from his dignity as a creature. These rights are prior to society and must be recognized by it. They are the basis of the moral legitimacy of every authority: by flouting them, or refusing to recognize them in its positive legislation, a society undermines its own moral legitimacy. If it does not respect them, authority can rely only on force or violence to obtain obedience from its subjects. It is the Church's role to remind men of good will of these rights and to distinguish them from unwarranted or false claims.
> Respect for the human person proceeds by way of respect for the principle that "everyone should look upon his neighbour (without any exception) as 'another self,' above all bearing in mind his life and the means necessary for living it with dignity." No legislation could by itself do away with the fears, prejudices, and attitudes of pride and selfishness which obstruct the establishment of truly fraternal societies. Such behaviour will cease only through the charity that finds in every man a "neighbour", a brother.
> The duty of making oneself a neighbour to others and actively serving them becomes even more urgent when it involves the disadvantaged, in whatever area this may be. "As you did it to one of the least of these my brethren, you did it to me."
> This same duty extends to those who think or act differently from us. The teaching of Christ goes so far as to require the forgiveness of offenses. He extends the commandment of love, which is that of the New Law, to all enemies. Liberation in the

> spirit of the Gospel is incompatible with hatred of one's enemy as a person, but not with hatred of the evil that he does as an enemy.[153]

Simply put, it is just not possible to be a Christian and ignore the plight of your neighbour, ignore the way we treat planet earth and all that live on her, and ignore social policy which is detrimental to human rights and to the full flourishing of the human being. Purveyors of evil may not like what the Church teaches in matters of both faith and morals, and may well affect to be "offended" by it. But that cannot deter the Church in her witness to the truth, and especially when those observations come from the Magisterium of the Catholic Church. The Church proposes, but does not impose. Secularists both propose and impose by their manipulation of Parliaments to construct laws which inhibit the free practice of religion and the freedom to fully proclaim the Catholic Faith.

Conclusion

In his encyclical *Laudato Si'*, Pope Francis does not use the traditional language of natural law He chooses to use a language of broader appeal, but it is easy to see from the authorities he cites that the teaching he reaffirms and develops is based on both Scripture and the natural law. I would agree with what *Catholic Earthcare Australia* says about this non-use of the traditional language of natural law, a non-use which, I think, is more apparent than real.

> What we see in LS is not a theological shift but an attempt to find new language for a broader audience. In this case, even those who don't have an ethos based on natural law can see that taking care of the environment for future generations is the right thing to do.[154]

And it is, I think, manifestly clear that care for the environment, which includes care for human beings, is a duty for us all deriving from the principles of the natural law which we can take for granted lies behind the Holy Father's approach to moral imperatives.

---

153 *Catechism of the Catholic Church*, nn. 1929-1933.
154 Catholic Earthcare Australia, *Questions and Answers dealing explicitly with the Encyclical*, n. 10 at http://catholicearthcare.org.au/wp-content/uploads/2015/05/Encyclical-Questions-and-Answers.pdf

Chapter 4

Catholic Social Teaching and the Environment

Introduction

Pope Francis put his signature to his Encyclical Letter *Laudato Si'* on the 24th of May 2015. It was more fully titled as *Laudato Si': On Care for our Common Home*.

In this chapter I will rely on the English edition to be found on the Vatican website.[155]

This Encyclical repeatedly calls for an open and honest debate on the topics addressed.[156] Those topics include certain aspects of Catholic social teaching, and a good deal of material touching on scientific, political, economic, and sociological matters. Given that, in the Australian edition, the Encyclical was misleadingly labelled *An Encyclical Letter on Ecology and Climate Change* Australians could be forgiven for thinking that this is all that the Encyclical is about.

To properly understand what is in the Encyclical the reader needs, however, to have some background knowledge of existing Catholic teaching on environmental questions, certain moral questions such as abortion, sterilisation, contraception and the like, and the limits of papal teaching where judgements are made involving disputed matters which do not belong to the provenance of faith and morals.

---

155 https://w2.vatican.va/content/dam/francesco/pdf/encyclicals/documents/papa-francesco_20150524_enciclica-laudato-si_en.pdf
156 *Laudato Si'*, nn. 16, 61, 135, 138, 188.

This Encyclical is addressed to the widest possible audience, to "every person living on this planet"[157] no matter what may be that person's attitude toward the Church, religion, or even good or evil! Where Christians are concerned, our fault is two-fold. In the first place many Christians seem to be completely unaware of what the Bible has to say about the environment, and how that material has been systematised and clarified by the Magisterium of the Catholic Church. And secondly, even when we are more or less aware of that teaching, we nevertheless choose, out of perceived self-interest, to sin anyway.

My intention in this chapter is firstly to outline the essential elements of Catholic social teaching where environmental issues are concerned prior to the issuing of *Laudato Si'*. Secondly, I will confirm the way in which Pope Francis reasserts Catholic social teaching, maintaining and furthering that teaching in relation to the state of affairs confronted by the world in the twenty first century. Thirdly I will explain how Catholic doctrine develops legitimately over time, and finally make some preliminary observations about *Laudato Si'*.

Catholic teaching on environmental issues prior to *Laudato Si'*

The tradition of Catholic social teaching offers a developing and distinctive perspective on environmental issues. That tradition rests upon the Church's foundational doctrines concerning God the creator, man[158] created by God in His image, and the rest of the created order committed to the care of human beings.

---

157 *Laudato Si'*, n. 3.
158 I am using the word "man" here in its traditional and inclusive sense, ie humankind both male and female.

God the creator

The first point of departure where Catholic social teaching is concerned is in the acknowledgment of the existence of God, the supreme and uncreated Being, who created everything. It is not within the ambit of this book to lay out the arguments for the existence of God. But the Church holds to and believes in the reality of God, the creator of all that is. She does so for two reasons:

> 1. through the application of human reason a person can come to know that God exists; and
> 2. that correct knowledge about what God is really like is revealed in Jesus Christ as described in both Holy Scripture and in the Tradition handed down by his closest followers, the apostles, who heard what he had to say and were eye witnesses to what he did.

The *Catechism of the Catholic Church* expresses it in this way:

> Human intelligence is surely already capable of finding a response to the question of origins. The existence of God the Creator can be known with certainty through his works, by the light of human reason, even if this knowledge is often obscured and disfigured by error. This is why faith comes to confirm and enlighten reason in the correct understanding of this truth: "By faith we understand that the world was created by the Word of God, so that what is seen was made out of things which do not appear." (Hebrews 11:3)[159]

And:

> "Sacred Scripture is the speech of God as it is put down in writing under the breath of the Holy Spirit." (*Dei verbum*, n. 9)
>
> "And [Holy] Tradition transmits in its entirety the Word of God which has been entrusted to the apostles by Christ the Lord and the Holy Spirit. It transmits it to the successors of the apostles so that, enlightened by the Spirit of truth, they may faithfully preserve, expound and spread it abroad by their preaching." (*Dei verbum*, n. 9)

---

159 *Catechism of the Catholic Church*, 286.

> As a result the Church, to whom the transmission and interpretation of Revelation is entrusted, "does not derive her certainty about all revealed truths from the holy Scriptures alone. Both Scripture and Tradition must be accepted and honoured with equal sentiments of devotion and reverence." (*Dei verbum*, n. 9)[160]

Next, the Church also holds that human beings, made in the image and likeness of God, know in their heart of hearts of their need for friendship with God, and that they have a responsibility to acknowledge that they need to behave in a way consistent with that friendship. The need for faith in God is part of the human experience. The Fathers of the Second Vatican Council expressed these propositions in this way:

> The dignity of man rests above all on the fact that he is called to communion with God. This invitation to converse with God is addressed to man as soon as he comes into being. For if man exists it is because God has created him through love, and through love continues to hold him in existence. He cannot live fully according to truth unless he freely acknowledges that love and entrusts himself to his creator.[161]

Moreover, the faculty of reason can only take us so far in knowing what God is really like. That being the case God chose to reveal Himself over time first to the Jewish people, and then ultimately to the whole world through Jesus Christ who is God and Man.

> By natural reason man can know God with certainty, on the basis of his works. But there is another order of knowledge, which man cannot possibly arrive at by his own powers: the order of divine Revelation. Through an utterly free decision, God has revealed himself and given himself to man. This he does by revealing the mystery, his plan of loving goodness, formed from all eternity in Christ, for the benefit of all men. God has fully revealed this plan by sending us his beloved Son, our Lord Jesus Christ, and

---

160 *Catechism of the Catholic Church*, nn. 81-82.
161 The Second Vatican Council, *Gaudium et spes*, n. 19.

the Holy Spirit.[162]

So then, on the basis of reason and the faith we rightly have in Jesus Christ who has certainly revealed God for who *He Is*, the Church reflects on what that revelation means in terms of the environment and man's responsibility as a steward. So a summary of Catholic social teaching where the environment is concerned would have regard to the following:

- Creation as a whole discloses to man the certainties a) that God exists, b) that God created all that is, and c) that the ultimate purpose and meaning of creation is to be found in the nature of God Himself.[163]
- Man is made in the image and likeness of God (*imago Dei*).[164]
- God created everything for man, but man in turn was created to serve and love God and to offer all creation back to him.[165]
- What is man that you are mindful of him?[166] Man has a special relationship with God and God a special care for man because God made man in His image and likeness. .
- Man is given dominion over the animals and over the earth, but that relationship of dominion is to be characterised by solidarity and benevolence as well as control.[167]
- Man is made as male and female. They complement each other in friendship and love. And they become one flesh as they share their conjugal love.[168] Through that love they are to "be fruitful and multiply".[169] In obedience to that command Adam and Eve had two sons, Cain and Abel.[170]
- The relationship between human beings is to be governed by mutual love and respect (from which virtues we must behave

---

162 CCC, 50.
163 Cf, CCC, 46.
164 Genesis 1:26.
165 Cf CCC, 358.
166 Psalm 8:4.
167 Cf CCC, 358.
168 Genesis 2:23-24.
169 Genesis 1:28.
170 Cf also CCC, 372.

justly, and more than justly).¹⁷¹ Homicide (ie the killings of the innocent) is everywhere condemned. Cain is condemned for killing his brother, Abel. But God also forbids the retributive killing of Cain, even though he is guilty of fratricide.¹⁷²

- After Abel's murder, Adam and Eve had another son, Seth, whom Eve believed was appointed by God as a replacement for Abel.¹⁷³
- After Seth, Adam and Eve went on to have more children, sons and daughters.¹⁷⁴

The relationship of man to animals and plants

The various Biblical references made here, and how we are to understand them, are to be found *passim* in the *Compendium of the Social Doctrine of the Church*:¹⁷⁵

- Since human beings are made in the image and likeness of God they have rights and responsibilities.
- Animals too are made out of the dust of the ground¹⁷⁶ and are also "living creatures".¹⁷⁷
- Animals are seen as friends and helpers of human beings¹⁷⁸, and are also called to "be fruitful and multiply"¹⁷⁹. So there is solidarity between human beings and animals. Man has stewardship over the animals because he is authorised to name them.¹⁸⁰

---

171 Matthew 5:40.
172 Genesis 4:8-16.
173 Genesis 4:25. "Later on, after Adam had sexual relations with his wife, she gave birth to a son and named him Seth, because "God granted me another offspring to replace Abel, since Cain murdered him."
174 Genesis 5:4. "Adam lived another 800 years, fathering other sons and daughters after he had fathered Seth."
175 Pontifical Council for Justice and Peace, *Compendium of the Social Doctrine of the Church*, Libreria Editrice Vaticana, 2004.
176 Genesis 2:19.
177 ibid.
178 Genesis 2:18.
179 Genesis 1:22.
180 Genesis 2:19-20.

- God saw all that he had made and saw it as good.[181]
- The creation is always seen as an object for which God is to be praised. "O Lord how manifold are your works! In wisdom you have made them all; the earth is full of your creatures."[182]
- This solidarity with animals (living beings) is specified in the Ten Commandments, where the Sabbath rest extends to "your cattle".[183] In the book of Deuteronomy this is further specified thus: "you shall not do any work...or your ox, or your ass, or any of your cattle."[184]
- In the Genesis story, after the Fall, the relationship between humans, between humans and animals, and humans and plants becomes less certain. Wild animals attack, droughts occur, and crops fail. The Old Testament provides a base line compromise with how human beings are to get on. After the flood they are given permission to eat animals but must not consume blood because to do so would be to disrespect life.[185]
- The Old Testament approach to animals is summed up in Proverbs 12: 10: "A righteous man has regard for the life of his beast, but the mercy of the wicked is cruel."
- Where care for the land is concerned, there is a requirement that the land be left fallow after six years and not cultivated in the seventh year. Here, the land is left to recover, what grows naturally being left to the poor and to wild animals. This looks like an ecological rule to help the land recover from its "work" while at the same time giving free reign to the poor[186]. Wild animals are not forgotten, they too are seen as living beings for whom human beings should have a care, despite the dangers to human beings from those animals.

The US Catholic Bishops Conference has summarised Catholic

---

181 Genesis 1:4, 10, 12, 18, 21, 25.
182 Psalm 104:24.
183 Exodus 20:10.
184 Deuteronomy 5:14.
185 Genesis 9:3-5.
186 For obligations to the poor in the stewardship of the created order see also Leviticus 19: 9-10; Deuteronomy 23: 24-25; 24: 19-22.

Social Teaching in relation to the integral dimensions of ecological responsibility as follows:

- a God-centred and sacramental[187] view of the universe, which grounds human accountability for the fate of the earth;
- consistent respect for human life, which extends to respect for all creation;
- a world view affirming the ethical significance of global interdependence and the common good;
- an ethics of solidarity promoting cooperation and a just structure of sharing in the world community;
- an understanding of the universal purpose of created things, which requires equitable use of the earth's resources;
- an option for the poor[188], which gives passion to the quest for an equitable and sustainable world;
- a conception of authentic development, which offers a direction for progress that respects human dignity and the limits of material growth.[189]

---

[187] "A thing may be called a "sacrament," either from having a certain hidden sanctity, and in this sense a sacrament is a "sacred secret"; or from having some relationship to this sanctity, which relationship may be that of a cause, or of a sign or of any other relation." STh III, q. 60, a. 1.

[188] *Centesimus annus*, n. 57: "Today more than ever, the Church is aware that her social message will gain credibility more immediately from the witness of actions than as a result of its internal logic and consistency. This awareness is also a source of her preferential option for the poor, which is never exclusive or discriminatory towards other groups. This option is not limited to material poverty, since it is well known that there are many other forms of poverty, especially in modern society—not only economic but cultural and spiritual poverty as well. The Church's love for the poor, which is essential for her and a part of her constant tradition, impels her to give attention to a world in which poverty is threatening to assume massive proportions in spite of technological and economic progress. In the countries of the West, different forms of poverty are being experienced by groups which live on the margins of society, by the elderly and the sick, by the victims of consumerism, and even more immediately by so many refugees and migrants. In the developing countries, tragic crises loom on the horizon unless internationally coordinated measures are taken before it is too late."

[189] United States Catholic Conference, *Catholic Social Teaching and Environmental Ethics*, http://www.webofcreation.org/DenominationalStatements/catholic.htm

"Sustainable world"

In the context of this discussion sustainability refers to the ability or capacity of something to maintain itself or to sustain itself well into the future. Sustainable means: "conserving an ecological balance by avoiding depletion of natural resources".[190] So sustainability refers to the good things which human beings need to live, and how these goods are to be obtained in ways that do not threaten subsequent generations' right to access these goods. It is simply the description of the goal to be achieved. By itself the goal does not instruct human beings as to what morally appropriate measures may be undertaken to achieve that goal.

In Catholic social teaching, the term "sustainable world" is not used in such a way that, important and all as the world is, it would suggest human existence on earth is man's ultimate end. The created world, too, has its end which is not entirely disconnected from the end of human beings. Man's end is to know God and to love Him, and to be able to enjoy his presence for eternity.

> The desire for God is written in the human heart, because man is created by God and for God; and God never ceases to draw man to himself. Only in God will he find the truth and happiness he never stops searching for.[191]

But this does not mean that man is or can be indifferent to the present world which has been given to us in trust. As I have remarked earlier, the stewardship of the world proper to human beings, requires of us respect for all that God has given us. Immoral policies designed to achieve a secular understanding of "sustainability", such as abortion, infanticide, contraception and the like, destroy our friendship with God and rule out our final and eternal destiny to be with God.

---

190 *Concise Oxford English Dictionary*, 2011.
191 *Catechism of the Catholic Church*, n. 27.

Marriage and the Family – the heart of civilisation

At the heart of Catholic social teaching on environmental issues is the good of the family based upon the union of a man and a woman in marriage. This is because the family is the natural and fundamental group unit of society. Families have formed societies in order that societies can serve families. Put another way, the State exists to serve families.[192]

The environmental challenges that confront humanity are, apart from natural causes, often the result of human acts and omissions. The solutions suggested in influential quarters centre around the idea that the human population on earth is excessive, is a cause in itself of environmental degradation, and that a reduction in population is not only desirable but necessary. Advocate for population control, Jeffrey Sachs, puts it this way:

> Higher energy use is already changing the world's climate in dangerous ways. Furthermore, the strains of increased global populations, combined with income growth, are leading to rapid deforestation, depletion of fisheries, land degradation, and the loss of habitat and extinction of a vast number of animal and plant species.
>
> Population growth in developing regions - especially Africa, India, and other parts of Asia - needs to slow. Public policies can play an important role by extending access to family planning services to the poor, expanding social security systems, reducing child mortality through public health investments, and improving education and job opportunities for women.[193]

By contrast, Saint John Paul II said this:

---

[192] Pope John Paul II describes the teaching of Pope Leo XIII: "On the contrary, he frequently insists on necessary limits to the State's intervention and on its instrumental character, inasmuch as the individual, the family and society are prior to the State, and inasmuch as the State exists in order to protect their rights and not stifle them." Cf *Centesimus annus*, n. 11, and Pope Leo XIII, *Rerum Novarum*, nn. 101f.; 104f.; 130f.; 136.

[193] Jeffrey Sachs, *The Case for Slowing Population Growth*, https://www.project-syndicate.org/commentary/the-case-for-slowing-population-growth

> In defense of the human person, the church stands opposed to the imposition of limits on family size and to the promotion of methods of limiting births which separate the unitive and procreative dimensions of marital intercourse, which are contrary to the moral law inscribed on the human heart or which constitute an assault on the sacredness of life. Thus sterilization, which is more and more promoted as a method of family planning, because of its finality and its potential for the violation of human rights, especially of women, is clearly unacceptable; it poses a most grave threat to human dignity and liberty when promoted as part of a population policy. Abortion, which destroys existing human life, is a heinous evil, and it is never an acceptable method of family planning, as was recognized by consensus at the Mexico City U.N. International Conference on Population (1984).[194] [Emphasis added]

Thirteen years later, Saint John Paul II directed our attention to the issues that are fundamental to understanding the moral position adopted by the population controllers.

> Today we often witness the taking of opposite and exaggerated positions: on the one hand, in the name of the exhaustibility and insufficiency of environmental resources, demands are made to limit the birth rate, especially among the poor and developing peoples. On the other, in the name of an idea inspired by egocentrism and biocentrism it is being proposed that the ontological and axiological difference between men and other living beings be eliminated, since the biosphere is considered a biotic unity of indifferentiated (*sic*) value. Thus man's superior responsibility can be eliminated in favour of an egalitarian consideration of the "dignity" of all living beings.
>
> But the balance of the ecosystem and the defence of the healthiness of the environment really need human responsibility and a responsibility that must be open to new forms of

---

[194] Talk given during a March 18 1994 meeting at the Vatican with Nafis Sadik, executive director of the U.N. Fund for Population Activities. The Pope, Saint John Paul II, criticised a final draft document prepared for the September U.N. International Conference on Population and Development in Cairo, Egypt.

> solidarity. An open and comprehensive solidarity with all men and all peoples is essential, founded on respect for life and the promotion of sufficient resources for the poorest and for future generations.[195] [Emphasis added]

The Catholic Church is sensitive to the urgent need to protect the environment because the creation is a gift from God and because human beings have been given the responsibility of doing everything scientifically and morally possible to be good stewards of this gift. There is a keen desire on the part of the Catholic Church to support all measures necessary to protect and enhance the natural environment.

> The balance of the ecosystem and the defence of the healthiness of the environment really need human responsibility and a responsibility that must be open to new forms of solidarity.[196]

But the Church does not take a one-sided view of environmental issues, and nor does she adopt a utilitarian approach to the solutions to environmental degradation. One must not only achieve a good end, but achieve it without violating other goods, such as the goods of natural marriage, procreation, and the family. It is never morally licit to do evil that good may come.[197] Evils may be different in kind, but evil is still evil, and the "disorder" we experience in the world has its origin in sin.[198]

> From the moral point of view contraception and abortion are specifically different evils. The former contradicts the full truth of the sexual act as the proper expression of conjugal love, while the latter destroys the life of a human being; the former is opposed to the virtue of chastity in marriage, the latter is opposed to the virtue of justice and directly violates the divine commandment "You shall not kill".[199]

---

195 Pope John Paul II, *Address to Conference on Environment and Health*, 24 March 1997, n. 5.
196 *Ibid.*
197 *Romans*, 3:8; and cf *Humanae vitae*, footnote 18.
198 Cf *Catechism of the Catholic Church*, nn. 403, 1607.
199 Pope John Paul II, *Evangelium vitae*, n. 13.

In fact, "abortion and infanticide are abominable crimes".[200]

Lynn White Jr's argument that science developed within the context of Christian civilisation is almost certainly true. But the Enlightenment represented a departure from orthodox Christianity such that, in the end, God is simply side-lined and ultimately retired from serious consideration in the way in which society developed. The temptation of "fallen humanity" to treat the rest of the created order as resources to be exploited for human convenience takes over from the Biblical insistence on the right order of human beings within the totality of the created order, and with a mandate to be good stewards of the earth. The Biblical data amply demonstrates the wrongheadedness of White on this important matter, and his promotion of a crude caricature of the Biblical evidence.

What White does get more or less right is when he observes that environmental degradation has emerged as a consequence of science and technology.

> But, as we now recognize, somewhat over a century ago science and technology – hitherto quite separate activities – joined to give mankind powers which, to judge by many of the ecologic effects, are out of control.[201]

But he does not in any way establish the truth of his subsequent conclusion that for this "Christianity bears a huge burden of guilt".

White is also right when he says:

> The victory of Christianity over paganism was the greatest psychic revolution in the history of our culture. It has become fashionable today to say that, for better or worse, we live in the "post-Christian age." Certainly the forms of our thinking and language have largely ceased to be Christian, but to my eye the substance often remains amazingly akin to that of the past. Our daily habits of action, for example, are dominated by an implicit

---

200 *Gaudium et spes*, n. 51 §3.
201 Lynn White Jr, "The Historical Roots of Our Ecological Crisis", *Ecology and Religion In History*, New York, Harper and row, 1974, 5.

faith in perpetual progress which was unknown either to Greco-Roman antiquity or to the Orient.²⁰²

But he is wrong when he then concludes that this "faith in perpetual progress" is "rooted in, and is indefensible apart from, Judeo-Christian theology".²⁰³ The reality is that the faith in progress derives from the Enlightenment thinking from Bacon, Hobbes, Locke, and the nineteenth century utilitarians.²⁰⁴

Pope Francis reasserts Catholic social teaching

Pope Francis, in *Laudato Si'*, restates and elaborates the scriptural basis of Catholic social teaching. In chapter two of this Encyclical, the Holy Father begins by acknowledging that some people deny the reality of God, or dismiss religion as irrelevant to discussions on the environment. But when they do this they miss out on the "rich contribution which religion can make towards an integral humanity."²⁰⁵ Nevertheless, the Pope insists that the different approaches to understanding reality in science and religion involve no necessary contradiction. Indeed these distinctive approaches can, he says, provide opportunity for "an intense dialogue fruitful for both".²⁰⁶

Indeed, as Pope Francis insists, "If we are truly concerned to develop an ecology capable of remedying the damage we have done, no branch of the sciences and no form of wisdom can be left out, and that includes religion and the language particular to

---

202 Lynn White Jr, "The Historical Roots of Our Ecological Crisis", *Ecology and Religion In History*, New York, Harper and row, 1974, 4. A copy of this paper may be found at https://www.uvm.edu/~gflomenh/ENV-NGO-PA395/articles/Lynn-White.pdf

203 *Ibid.*

204 This is not the place to trace the history of ideas, but suffice to say that Hobbes' hedonic principle, viz that the good is that which gives pleasure and the evil that which causes pain, was later taken up by Bentham and Mill to develop the highly improbable idea that pleasures can be added up, pains added up, and the one subtracted from the other to guide us as to what is the morally right thing to do.

205 *Laudato Si'*, n. 62.

206 *Ibid.*

it."²⁰⁷ The particular genius of the Catholic Church stems from her synthesis of faith and reason which is especially evident in her social teaching.²⁰⁸

So in this contribution to Catholic social teaching, set out especially in chapter 2, Pope Francis refers to all of the major issues at stake where environmental matters are concerned.

The Book of Genesis

The Encyclical positions the creation stories in Genesis in their rightful context, drawing attention to its "profound teachings about human existence and its historical reality" in language with is both narrative and symbolic. These stories:

> ... suggest that human life is grounded in three fundamental and closely intertwined relationships: with God, with our neighbour and with the earth itself. According to the Bible, these three vital relationships have been broken, both outwardly and within us. This rupture is sin. The harmony between the Creator, humanity and creation as a whole was disrupted by our presuming to take the place of God and refusing to acknowledge our creaturely limitations. This in turn distorted our mandate to "have dominion" over the earth (cf. Gen 1:28), to "till it and keep it" (Gen 2:15).²⁰⁹

This "presuming to take the place of God" is clearly evident in Enlightenment and post-Enlightenment thinking as described above. But, "we are not God"!²¹⁰ The earth was here long before human beings came on the scene. At this point the Pope deals strongly with the claim made by Lynn White Jr and others, that Genesis has "encouraged the unbridled exploitation of nature by painting him as domineering and destructive by nature." Acknowledging that Christians have at times "incorrectly interpreted the Scriptures, nowadays we must forcefully reject the notion that our being created

---

207 *Ibid.*, n. 63.
208 *Ibid.*
209 *Ibid.*, n. 66.
210 *Ibid.*, n. 67.

in God's image and given dominion over the earth justifies absolute domination over other creatures."[211]

The Pope then details the references in the books of the Old Testament which command human beings to "till and keep" the earth. Yes, the earth may be used for cultivating the produce we need to eat, a process which is one of cooperation between the farmer and the soil. And yes, we must "keep" the earth, protect it to ensure its continuing fruitfulness. But we don't own the earth. On the contrary, the Pope reminds us, "the earth is the Lord's" (Psalm 24:1").

> Thus God rejects every claim to absolute ownership: "The land shall not be sold in perpetuity, for the land is mine; for you are strangers and sojourners with me" (Lev 25:23).[212]

In the next few paragraphs the Pope rejects what he calls "tyrannical anthropocentrism" because it is inconsistent with what the Bible requires. He reminds us of the "delicate equilibria" which must exist between all the creatures of this world including man himself. He reminds us again of the respect we humans must have for animals, who are our companions. This sensitivity is well expressed in the books of Deuteronomy and Exodus:

> The laws found in the Bible dwell on relationships, not only among individuals but also with other living beings. "You shall not see your brother's donkey or his ox fallen down by the way and withhold your help… If you chance to come upon a bird's nest in any tree or on the ground, with young ones or eggs and the mother sitting upon the young or upon the eggs; you shall not take the mother with the young" (Dt 22:4, 6). Along these same lines, rest on the seventh day is meant not only for human beings, but also so "that your ox and your donkey may have rest" (Ex 23:12).[213]

Clearly, the Pope states forcefully, "the Bible has no place for a tyrannical anthropocentrism unconcerned for other creatures."[214]

---

211 *Ibid.*
212 *Ibid.*
213 *Ibid.*, n. 68.
214 *Ibid.*

The value of other living beings

Other living creatures have value too. They are not to be mistreated, considered merely as fodder to satisfy human self-interest.

> By virtue of our unique dignity and our gift of intelligence, we are called to respect creation and its inherent laws, for "the Lord by wisdom founded the earth". (Proverbs 3:19)[215]

He then draws our attention to the way these teachings are expressed in the *Catechism of the Catholic Church*. The *Catechism*, he says, "clearly and forcefully criticizes a distorted anthropocentrism:

> Each creature possesses its own particular goodness and perfection... Each of the various creatures, willed in its own being, reflects in its own way a ray of God's infinite wisdom and goodness. Man must therefore respect the particular goodness of every creature, to avoid any disordered use of things.[216]

Human sin and environmental degradation

Who then is responsible for man-made environmental degradation? We all share responsibility for that. Why have we done to the environment what we have done? It is because of sin, due to our hubris, our demand that as individuals we should be free to exploit nature in any way that we wish and especially when we stand to benefit from it. Very early in this Encyclical Pope Francis cites the warnings given by three of his predecessors.

1. Blessed Pope Paul VI

> In 1971, eight years after *Pacem in Terris*, Blessed Pope Paul VI referred to the ecological concern as "a tragic consequence" of unchecked human activity: "Due to an ill-considered exploitation of nature, humanity runs the risk of destroying it and becoming in turn a victim of this degradation".[217]

---

215 *Ibid.*, n. 69.
216 CCC, n. 339, and cited in *Laudato Si'*, 9.
217 *Laudato Si'*, n. 4. The reference to Blessed Pope Paul VI comes from the Apostolic Letter *Octogesima Adveniens* (14 May 1971), 21: AAS 63 (1971), 416-417.

2. Saint John Paul II

Saint John Paul II became increasingly concerned about this issue. In his first Encyclical he warned that human beings frequently seem "to see no other meaning in their natural environment than what serves for immediate use and consumption". Subsequently, he would call for a global ecological conversion. At the same time, he noted that little effort had been made to "safeguard the moral conditions for an authentic human ecology". The destruction of the human environment is extremely serious, not only because God has entrusted the world to us men and women, but because human life is itself a gift which must be defended from various forms of debasement. Every effort to protect and improve our world entails profound changes in "lifestyles, models of production and consumption, and the established structures of power which today govern societies". Authentic human development has a moral character. It presumes full respect for the human person, but it must also be concerned for the world around us and "take into account the nature of each being and of its mutual connection in an ordered system". Accordingly, our human ability to transform reality must proceed in line with God's original gift of all that is.[218]

3. Pope Benedict XVI

My predecessor Benedict XVI likewise proposed "eliminating the structural causes of the dysfunctions of the world economy and correcting models of growth which have proved incapable of ensuring respect for the environment". He observed that the world cannot be analyzed by isolating only one of its aspects, since "the book of nature is one and indivisible", and includes the environment, life, sexuality, the family, social relations, and so forth. It follows that "the deterioration of nature is closely

---

[218] *Laudato Si'*, n. 5. The references to Saint John Paul II are from *Redemptor Hominis* (4 March 1979), 15, 287; Catechesis (17 January 2001), 4; *Insegnamenti* 41/1 (2001); *Centesimus Annus* (1 May 1991), 38 and 37, 841; and *Sollicitudo Rei Socialis* (30 December 1987), 34, 559.

connected to the culture which shapes human coexistence". Pope Benedict asked us to recognize that the natural environment has been gravely damaged by our irresponsible behaviour. The social environment has also suffered damage. Both are ultimately due to the same evil: the notion that there are no indisputable truths to guide our lives, and hence human freedom is limitless. We have forgotten that "man is not only a freedom which he creates for himself. Man does not create himself. He is spirit and will, but also nature". With paternal concern, Benedict urged us to realize that creation is harmed "where we ourselves have the final word, where everything is simply our property and we use it for ourselves alone. The misuse of creation begins when we no longer recognize any higher instance than ourselves, when we see nothing else but ourselves".[219]

So Pope Francis strongly reasserts Catholic social teaching on the environment, reminding us again that the whole of humanity has been in one way or another complicit in the causes of much environmental degradation, that man must stop carrying on as if he is all that really matters and can do as he pleases with the world in which he lives, that man does not "own" the created order, that man is part of nature, that continuing disrespect for life, human and non-human, can only bring us misery; and that this human narcissism, what the Pope calls "tyrannical anthropocentrism", has to be controlled if we truly wish to allow the planet to recover from the mistreatment it has received at our hands.

The moral plane that needs to be restored to human societies is that which finds its foundations in the natural moral law enlightened by the Light of Christ. This applies to us all, Christians, those of other faiths, and those of no religion.

> The best way to restore men and women to their rightful place, putting an end to their claim to absolute dominion over the earth,

---

219 *Laudato Si'*, n. 6. The references to Pope Benedict XVI are from *Address to the Diplomatic Corps Accredited to the Holy See* (8 January 2007), 73; *Caritas in Veritate* (29 June 2009), 51; *Address to the Bundestag,* Berlin (22 September 2011); and Address to the Clergy of the Diocese of Bolzano-Bressanone (6 August 2008).

> is to speak once more of the figure of a Father who creates and who alone owns the world. Otherwise, human beings will always try to impose their own laws and interests on reality.[220]

Pope Francis acknowledges that the world is not perfect as it is, and that the world can be nurtured and developed to expose its full potential. God, he says, wants to cooperate with us, work with us, even to assist us to "bring good out of the evil we have done."[221] He then goes on to bring out from the *Catechism of the Catholic Church*, Catholic teaching on the existence of an imperfect world.

> But why did God not create a world so perfect that no evil could exist in it? With infinite power God could always create something better. But with infinite wisdom and goodness God freely willed to create a world "in a state of journeying" towards its ultimate perfection. In God's plan this process of becoming involves the appearance of certain beings and the disappearance of others, the existence of the more perfect alongside the less perfect, both constructive and destructive forces of nature. With physical good there exists also physical evil as long as creation has not reached perfection.[222]

So we have a world which needs development, but that development needs to be carried out on a morally sound basis. Another human being is a subject just as I am and can "never be reduced to the status of an object."[223] And nor are other living beings to be considered as "mere objects subjected to arbitrary human domination".

> When nature is viewed solely as a source of profit and gain, this has serious consequences for society. This vision of "might is right" has engendered immense inequality, injustice and acts of violence against the majority of humanity, since resources end up in the hands of the first comer or the most powerful: the winner takes all. Completely at odds with this model are the ideals of

---

220 *Laudato Si'*, n. 75.
221 *Laudato Si'*, n. 80.
222 CCC, 310. This paragraph further references St Thomas Aquinas, *STh* I, 25,6; St Thomas Aquinas, *SCG III*, 71.
223 *Laudato Si'*, n. 81.

harmony, justice, fraternity and peace as proposed by Jesus. As he said of the powers of his own age: "You know that the rulers of the Gentiles lord it over them, and their great men exercise authority over them. It shall not be so among you; but whoever would be great among you must be your servant" (Mt 20:25-26).[224]

Integral Ecology in *Laudato Si'*

I have already drawn attention to the rejection by Pope Francis of "tyrannical anthropocentrism". Again drawing from the whole Catholic tradition, the Holy Father reaffirms the complex interdependence of living things. Human beings need to see themselves "in relation to all other living creatures".[225]

There has always been a tendency among some scientists to disrespect what they don't understand. The description of 90% of human DNA as "junk DNA" is, to say the very least, unfortunate. Now that that description has been challenged what has followed has been an unseemly debate where partisans of the "junk DNA" proposition hurl scorn on the scientists from Endcode who now suggest that 80% of DNA is functional. It is not for me to enter the debate as to who is right and who is wrong[226], but simply to observe that the unfortunate choice of the word "junk" epitomises man's hubris where the created order is concerned.

Another example of dismissing what we don't understand is the "appendix" in the human body, long regarded as serving no useful purpose. But in 2013 researchers "discovered the true function of this organ, and it is anything but redundant. Researchers now say that the appendix acts as a safe house for good bacteria. The body uses this to essentially "reboot" the digestive system when one

---

224 *Ibid.*, n. 82.
225 *Ibid.*, n. 85.
226 A useful discussion of the issues at stake may be found in Philip Ball, "Is junk DNA all garbage?" 22 May 2014, http://www.rsc.org/chemistryworld/2014/05/junk-dna-all-garbage

suffers from a bout of dysentery or cholera."[227]

Extend this to a list of "creepy crawlies" which we think could safely be exterminated until we find their contribution to the ecology of the world considered as a whole. Even cockroaches have their place and their importance in the ecosystem!

But the longstanding position of Catholic theologians is that the interdependence of living organisms has to be protected. Pope Francis calls to his witness the Angelic Doctor, Saint Thomas Aquinas, in his elegantly phrased paragraph 86 which extols the wonders of the universe.

> The universe as a whole, in all its manifold relationships, shows forth the inexhaustible riches of God. Saint Thomas Aquinas wisely noted that multiplicity and variety "come from the intention of the first agent" who willed that "what was wanting to one in the representation of the divine goodness might be supplied by another",[228] inasmuch as God's goodness "could not be represented fittingly by any one creature".[229] Hence we need to grasp the variety of things in their multiple relationships.[230] We understand better the importance and meaning of each creature if we contemplate it within the entirety of God's plan. As the Catechism teaches: "God wills the interdependence of creatures. The sun and the moon, the cedar and the little flower, the eagle and the sparrow: the spectacle of their countless diversities and inequalities tells us that no creature is self-sufficient. Creatures exist only in dependence on each other, to complete each other, in the service of each other".[231]

Of course Pope Francis has no wish to "put all living beings on the same level nor to deprive human beings of their unique worth and

---

[227] MB David, "Scientists Finally Discover the Function of the Human Appendix", Jul 22, 2013, http://politicalblindspot.com/scientists-finally-discover-the-function-of-the-human-appendix/

[228] Saint Thomas Aquinas, *Summa Theologiae*, I, q. 47, art. 1.

[229] *Ibid.*

[230] Cf. *ibid.*, art. 2, ad 1; art. 3.

[231] CCC, 340

the tremendous responsibility it entails."[232] He holds in tension the unique place of human beings, created in the image of God, with the rest of the created order animate and inanimate. And nor does the Pope want us to make the leap towards pantheism whereby nature becomes invested with a Divine reality which it does not have. No, we human beings are to work on the earth, "protecting it in its fragility".[233]

In Chapter Four of the Encyclical Pope Francis goes on to describe what he means by "integral ecology".

> Since everything is closely interrelated, and today's problems call for a vision capable of taking into account every aspect of the global crisis, I suggest that we now consider some elements of an *integral ecology*, one which clearly respects its human and social dimensions.[234]

Here the Holy Father again reasserts Catholic social teaching and further elaborates that teaching on this subject. He discusses this ecology in terms of environmental, economic and social concerns which are all part of the total ecology we live and experience in the total created order. Human beings, and their capacity for rational thought, are central, because we human beings can make choices which affect for good or ill the conditions of other human beings, and nature considered as a whole.

As he put it earlier in the Encyclical, at the centre of our concern for nature must be God, and "a sense of deep communion with the rest of nature". That sense of deep communion will not be real, "if our hearts lack tenderness, compassion and concern for our fellow human beings".[235] How inconsistent would it be if we are willing to combat "trafficking in endangered species" while at the same time remaining completely "indifferent to human trafficking"?[236]

---

232 *Laudato Si'*, n. 90.
233 *Ibid.*
234 *Ibid.*, n. 137.
235 *Ibid.*, n. 91.
236 *Ibid.*

Development of Catholic Doctrine

Pope Francis has entered into the global debate on the environment, calling for an even more urgent debate on environmental issues. True to the synthesis of faith and reason, Francis provides an alternative way of looking at the world and the place of man in it.

Each pope relies on past Church teachings and seeks to develop further a more complete understanding of those teachings. In this way the Church's teachings develop and maintain their freshness in the light of the circumstances of the modern world. But what does the "development of Catholic doctrine" entail? There are some who misunderstand the notion of "development of doctrine" to mean coming to the opposite conclusion to the original statement of the doctrine concerned. So how do we discern true development of doctrine and therefore understand how and why popes teach as they do?

As I say, in our time there are some who wrongly use the principle of "development of doctrine" to mean that the Church can change her dogmatic teaching in the light of developments in science and society, so that what was once true becomes now substantially untrue. Some of the same words may be used, but they are filled with a different content, content at odds with their original meaning. The origin of the principle of the "development of doctrine" may be found in the writings of Saint Vincent Lérins in the 5th century AD who said this:

> Therefore, let there be growth and abundant progress in understanding, knowledge, and wisdom, in each and all, in individuals and in the whole Church, at all times and in the progress of ages, but only with the proper limits, i.e., within the same dogma, the same meaning, the same judgment [eodem sensu eademque sententia].[237]

In other words, a development in doctrine is valid only when it takes place within the boundaries of the doctrine itself. It cannot and

---
237 *Commonitórium primum*, chapter 23, n. 54.

must not conclude or affirm the contrary of the original teaching. Saint Vincent then makes a crucial distinction between "progress" and "change", a distinction which seems to have been lost on many key figures in current debates within the Catholic Church.

> But it [progress of religion] must be such as may be truly a progress of the faith, not a change; for when each several thing is improved in itself, that is progress; but when a thing is turned out of one thing into another, that is change.[238]

This teaching of Saint Vincent Lérins was affirmed by the *Council of Trent* (1545-1563) and the *First Vatican Council* (1869-1870).[239] And on the 11th of October 1962 Pope Saint John XXIII also appealed to it in his opening address at the Second Vatican Council.[240]

However, at the Second Vatican Council (1962-5) defects in Saint Vincent's formulation of legitimate development of doctrine were acknowledged and a better approach adopted. One of the *periti* at Vatican II, Joseph Cardinal Ratzinger, explains the move away from the Vincentian canon.

> He [Lérins] no longer appears an authentic representative of the Catholic idea of tradition, but outlines a canon of tradition based on a semi-Pelagian idea. He attacks Augustine's teaching on grace as going beyond 'what had always been believed', but against this background this proves to be an inappropriate

---

238 *Ibid.*

239 *Dei filius, the Dogmatic Constitution on the Catholic Faith*, chapter 4, "Faith and Reason". For, the doctrine of faith which God revealed has not been handed down as a philosophic invention to the human mind to be perfected, but has been entrusted as a divine deposit to the Spouse of Christ, to be faithfully guarded and infallibly interpreted. Hence, also, that understanding of its sacred dogmas must be perpetually retained, which Holy Mother Church has once declared; and there must never be recession from that meaning under the specious name of a deeper understanding "Therefore [...] let the understanding, the knowledge, and wisdom of individuals as of all, of one man as of the whole Church, grow and progress strongly with the passage of the ages and the centuries; but let it be solely in its own genus, namely in the same dogma, with the same sense and the same understanding." [Vincent of Lerins, Commonitorium, 23, 3]."

240 *Gaudet Mater Ecclesia.* "The deposit or the truths of faith, contained in our sacred teaching, are one thing, while the mode in which they are enunciated, keeping the same meaning and the same judgment [eodem sensu eademque sententia], is another."

attempt to express the relationship between constancy and growth in the testimony of faith. The rejection of the suggestion to include again Vincent de Lérin's well known text, more or less canonized by two councils, is again a step beyond Trent and Vatican I…It is not that Vatican II is taking back what was intended in those quotations: the rejection of a modernistic evolutionism, an affirmation of the definitive character of the revelation of Christ and the apostolic tradition, to which the Church has nothing to add, but which is its yardstick, but it has another conception of the nature of historical identity and continuity. Vincent de Lérin's static *semper* no longer seems the right way of expressing the problem.[241]

Perhaps the most succinct formulation of this fresh approach to the development of doctrine may be found with reference to Blessed John Henry Newman's seven notes or tests of legitimate doctrinal development.[242]

The following formulations of Newman's tests are taken from Brendan Murphy, 'The Development of Doctrine: Is Catholic teaching a corruption of the "simple" Gospel?'[243]

Unity of Type

The first note of genuine development is unity of type. Newman considered this first criterion the most important of the seven. What he means by type is the external expression of an idea. The unity or preservation of type refers to the continual presence of a main idea despite its changing external expression. When we see change in the teaching on a subject, can we discern nevertheless that the

---

241 Joseph Ratzinger, 'The Transmission of Divine Revelation' in Herbert Vorgrimler, (ed), *Commentary on the Documents of Vatican II, Vol. III* (New York: Herder and Herder, 1969), 187. I am indebted to Professor Tracey Rowland, Dean of the John Paul II Institute for Marriage and family (Melbourne) for drawing my attention to this source.

242 See chapter 5 of John Henry Cardinal Newman, *An Essay on the Development of Christian Doctrine*, written in 1845.

243 OSV Newsweekly, https://www.osv.com/OSVNewsweekly/Story/TabId/2672/ArtMID/13567/ArticleID/9477/The-Development-of-Doctrine.aspx

main idea remains unchanged? If so, we know that the change is a genuine development, not a corruption.

Continuity of Principles
Newman insists that for a development to be faithful, it must preserve the principle with which it started. While doctrine may grow and develop, principles are permanent.

Power of Assimilation
In introducing this criterion, Newman notes that in the physical world living things are characterized by growth, not stagnancy, and that this growth comes about by making use of external things. For example, as human beings we grow by taking into our bodies external realities such as food, water and air.

In Newman's terminology, then, when we make use of these resources we are assimilating them. The food, water and air we consume don't change who or what we are in any meaningful way. Rather, they serve a valuable function in that they ensure our continued growth and vitality.

Logical Sequence
By this Newman means that a doctrine that's defined and professed by the Church at a point historically distant from its original founding can be considered a development, and not a corruption, if it can be shown to be the logical outcome of the original teaching. Newman compares this process to the growth of a tree. Someone looking at an oak tree could very easily draw the conclusion that it has nothing at all in common with an acorn. Yet the mature oak tree is the logical development of the acorn.

Anticipation of Its Future
The fifth note of genuine development may be seen as a corollary of the previous one.

Doctrines in some way imply or allude to their later development.

So authentic developments will have some logical connection to the original deposit of faith, however vague the "embryonic" form might have been in the earliest days of the Church.

Conservative Action
The sixth note of genuine development is conservative action upon its past.

A development is not a corruption if the doctrine proposed builds upon the doctrinal developments that precede it, often clarifying and strengthening them. A corrupt doctrine, on the other hand, is one that contradicts or reverses a preceding doctrinal development.

Chronic Vigour
The seventh note of genuine development is chronic – that is, abiding – vigour.

As long as a doctrine maintains its life and vigour, its ongoing development is assured. However, once a corruption enters into the process, it leads, by its nature, to death and decay.

In short, there can be no authentic development in Catholic doctrine where the proposed formulation fails these tests.

The social doctrine contained in this Encyclical clearly meets the requirements or tests of doctrinal development and continuity.

Laudato Si': some initial observations

Recalling briefly what was set out in chapter 1 of this book, an encyclical is a means by which the Pope addresses a specific group of persons, or as in the case of *Laudato Si'*, the whole planet. The kinds of things that popes have wished to communicate by this means have varied over the centuries.

And "the content and context" of the Encyclical has to be considered when interpreting the message. Catholic social teaching is derived from the Church's authority to teach on matters of faith and morals and in the case of a Papal Encyclical from the Pope's specific teaching authority. Even there, "the intention of the Pope in issuing the document" must also be considered together with "the way in which the bishops and the whole Church receive the teaching."

Moreover, a clear distinction should be made between the articulation of Catholic social teaching *per se* and the suggested ways in which that teaching should be applied prudentially in particular circumstances.

Laudato Si' makes no new infallible statements where Catholic social teaching is concerned, but it does do two main things. It reiterates and develops Church social teaching where the environment is concerned. And it applies that teaching to the situation in which the Pope believes we now find ourselves.

Reiterating Catholic social teaching on the environment, the Pope reinforces key aspects of social teaching in harmony with his predecessors. For example, against the population controllers who are committed to promoting, and in some cases enforcing, programmes of contraception, sterilisation and abortion, Pope Francis makes this forceful and magisterial response:

> Instead of resolving the problems of the poor and thinking of how the world can be different, some can only propose a reduction in the birth rate. At times, developing countries face forms of international pressure which make economic assistance contingent on certain policies of "reproductive health". Yet "while it is true that an unequal distribution of the population and of available resources creates obstacles to development and a sustainable use of the environment, it must nonetheless be recognized that demographic growth is fully compatible with an integral and shared development" [Footnote: Pontifical Council for Justice and Peace, *Compendium of the Social Doctrine of the Church*, 483]. To blame population growth instead of extreme and selective consumerism on the part of some, is one way of refusing to face the issues. ... Besides, we know that approximately a third of all food produced is discarded, and "whenever food is thrown out it is as if it were stolen from the table of the poor".[244]

---

[244] *Laudato Si'*, n. 50.

And Pope Francis again reiterates, strongly, the centrality of the family in all of our considerations about the environment.

> Ecological education can take place in a variety of settings: at school, in families, in the media, in catechesis and elsewhere. Good education plants seeds when we are young, and these continue to bear fruit throughout life. Here, though, I would stress the great importance of the family, which is "the place in which life – the gift of God – can be properly welcomed and protected against the many attacks to which it is exposed, and can develop in accordance with what constitutes authentic human growth. In the face of the so-called culture of death, the family is the heart of the culture of life". In the family we first learn how to show love and respect for life; we are taught the proper use of things, order and cleanliness, respect for the local ecosystem and care for all creatures. In the family we receive an integral education, which enables us to grow harmoniously in personal maturity. In the family we learn to ask without demanding, to say "thank you" as an expression of genuine gratitude for what we have been given, to control our aggressivity and greed, and to ask forgiveness when we have caused harm. These simple gestures of heartfelt courtesy help to create a culture of shared life and respect for our surroundings.[245]

So, as observed earlier, the social doctrine contained in this Encyclical clearly meets the requirements or tests of doctrinal development and continuity where the "care for our common home" is concerned. The great strength of *Laudato Si'* is the way that Pope Francis has brought together in one place so much of the tradition, further developing those crucial themes into one place, and presenting that teaching as gift to every person on the planet.

---

245 *Laudato Si'*, n. 213.

CHAPTER 5

Laudato Si': a contribution to "the need for forthright and honest debate"[246]

The reporting in secular journals of the debate on environmental degradation, its causes, and what is to be done has generally centred on the perspectives of secular humanist supporters of the "green movement", "green politics", and the secularists' own particular moral beliefs. This is part of the total political, social and cultural environment to which the Pope draws attention and which he addresses at certain points in his Encyclical.

It is abundantly clear that more often than not the values of those who have positioned themselves at the centre of the local, national, and international environmental debates are at odds with the agreed moral perceptions of the peoples of the world as expressed in the various fundamental human rights documents enunciated and proclaimed by the United Nations. Even within the United Nations itself the dominant mind-set of their highly paid operatives is in dissent from these human rights documents, especially when it comes to matters such as sex, birth control, family, contraception, sterilisation, abortion, and freedom. Examples of this abound but in this chapter I shall deal only with those issues relevant to environmental issues, and as they are presented by Jeffrey D Sachs, Special Advisor to the Secretary-General on the *Millennium Development Goals*, and Director of *The UN Millennium Project*.

---

246 *Laudato Si'*, n. 16.

A word of caution

Engagement in the political process is a social and civic duty. But it must proceed on a morally sound basis. That means that Catholics must be "wise as serpents and innocent as doves."[247] They must also be vigilant to discern when they are coming up against "false prophets, who come in sheep's clothing but inwardly are ravenous wolves."[248] Especially in the areas of threats to human life, we see the ubiquitous presence of urbane politically savvy experts, who hide their real agenda through the use of euphemisms. In this way they seek to seduce Catholics to naively accept, at face value, what they have to say.

We have already seen what the population controllers mean when they use terms such as "gender equality", "reproductive rights", and "sexual and reproductive health". These terms are code for abortion, contraception and sterilisation.

But even those in high places in the Church do not always seem to be aware of just who are the wolves in sheep's clothing when it comes to these issues. A recent example of this naivety is to be found in an interview given by Archbishop Marcelo Sánchez Sorondo, Chancellor of the *Pontifical Academies of Science and Social Sciences*, which hosted the "Protect the Earth, Dignify Humanity" conference at the Vatican in April 2015 in which Secretary-General of the United Nations, Ban Ki-moon, and his adviser Jeffrey D Sachs, were invited to participate.

> Q. Undoubtedly, you discussed Ban Ki-moon's and Jeffrey Sachs' position on abortion and population control in the lead up to the conference. How were any questions resolved?
> S.S. Yes. We had these discussions, and as you can see, the draft SDGs (Sustainable Development Goals) don't even mention abortion or population control. They speak of access to family planning and sexual and reproductive health and reproductive rights. The interpretation and application of these depends on

---
247 Matthew 10:16.
248 Matthew 7:15.

governments. Some may even interpret it as Paul VI, in terms of responsible paternity and maternity. Instead of attacking us, why not enter into dialogue with these "demons" to maybe make the formulation better, like we did on the issues of social inclusion and new forms of slavery?[249]

Such culpable naivety conspires to undermine, in a consistent way, application of magisterial teaching on population control, the environment, and essential elements of international human rights law. Here Church officials are seduced by accepting at face value seemingly innocuous language, but language which is used to promote destructive and morally bankrupt policies.

And to invoke the name of Pope Paul VI, suggesting that the Sustainable Development Goals can be interpreted in line with the teaching in *Humanae vitae* is risible. How could Pope Paul VI be in any way associated with the promotion of contraception, sterilisation and abortion?

Just how hard is it to read the real intentions of those who promote the language of "reproductive rights" and "sexual and reproductive health"? Those who invented these terms and use them are quite open about what they really mean as has already been explained in chapter 2 above. They mean that population growth should be curtailed, and that it should be curtailed by embracing policies which promote abortion, contraception, and sterilisation and which may even, in some circumstances be forcibly imposed.

Saint John Paul II, in his Encyclical, *Evangelium vitae*, spelled out the internal incoherency of 'population planner' policies and their manifest dangers which, while purporting to uphold the *Universal Declaration of Human Rights* (and related instruments), at the same time promote solutions to global problems which continue to devalue and destroy human lives.

> A society lacks solid foundations when, on the one hand, it asserts values such as the dignity of the person, justice and peace, but then, on the other hand, radically acts to the contrary by allowing or

---

249 https://c-fam.org/turtle_bay/vatican-prelate-blasts-critics-of-climate-conference/

tolerating a variety of ways in which human life is devalued and violated, especially where it is weak or marginalized. Only respect for life can be the foundation and guarantee of the most precious and essential goods of society, such as democracy and peace.[250]

The preferential option for human life

At the centre of Pope Francis' concerns is what I call the *preferential option for human life*. All that the Pope is on about is the good of human beings in their relationship with God, their relationship with their fellow human beings, and their relationship to the natural world in which human beings are integrally related.[251] Human beings have an inherent dignity, made as they are in the image and likeness of God. They are moral agents who can make free choices. Their freely made decisions affect their relationship with God, man, and the entire created order.

So it is that the dignity of the human being is enhanced when morally good and prudentially sound decisions are made, and damaged when morally bad and imprudent decisions are made. The good of human life characterised by what is called the inalienable and inviolable right to life of every human being (the *sine qua non* of all other rights) is front and centre in this encyclical.

Human beings are born into families, some good, some not so good, and some dysfunctional given the sin of the world.

Notwithstanding the fact that we live in a world marred by sin, the family, based upon natural marriage, is the natural and fundamental group unit of society, the original social unit upon which the state relies.[252] Man, made in the image of God, has been given the responsibility to look after the environment and to discern all morally sound ways to either prevent or repair environmental degradation.

Catholic social teaching provides both the basis and the ways

---

250 *Evangelium vitae*, n. 101.
251 Cf for example, *Laudato Si'* nn. 10, 42, 66, 90, 91, 137, 138.
252 *Universal Declaration on Human Rights*, Article 16(3); *International Covenant on Civil and Political Rights*, Article 23 (1).

to meet environmental crises caused by man's inhumanity to man, greed, carelessness, and indifference. Once these moral foundations are accepted, solutions to problems become more obvious. But we also must strive against the worst aspects of our fallen human nature. Confronted by major socio-moral problems of our own creating, the temptation is to find "easy" solutions, cruel solutions, solutions of the kind that got us into the mess in the first place. These kinds of solutions to man-made problems, in turn, create a raft of new problems which will further coarsen and degrade our moral sensibilities.

Pope Francis and the Enlightenment

In chapter two of this book I set out what I consider to be the ecological consequences of the working out of the philosophical underpinnings of the Enlightenment and post-Enlightenment periods. With man now at the centre of human history, God having been retired to the realms of private opinion, human hubris free from moral restraint takes over from prudential sanity. "If we can do it, why not do it?" became the operational principle where science and technology are concerned. "You can't put the genie back in the bottle" was frequently cited as one form of the technological imperative, never mind what damage the genie might do. It was as if we were helpless before the demands of science and technology. Pope Francis rejects this way of thinking. The applications of the knowledge and power which comes from advances in the sciences to new technologies, industry, and the like must be built upon a solid moral foundation. The Pope freely and joyfully recognises all of the benefits that have come to human beings over the last couple of centuries through the extraordinary advances in science and technology.

> We are the beneficiaries of two centuries of enormous waves of change: steam engines, railways, the telegraph, electricity, automobiles, aeroplanes, chemical industries, modern medicine, information technology and, more recently, the digital

revolution, robotics, biotechnologies and nanotechnologies. It is right to rejoice in these advances and to be excited by the immense possibilities which they continue to open up before us, for "science and technology are wonderful products of a God-given human creativity".[253]

But knowledge and the power that comes with it have to be used responsibly. And here the Pope sounds a serious note of warning:

> The fact is that "contemporary man has not been trained to use power well", because our immense technological development has not been accompanied by a development in human responsibility, values and conscience. Each age tends to have only a meagre awareness of its own limitations. It is possible that we do not grasp the gravity of the challenges now before us. "The risk is growing day by day that man will not use his power as he should"; in effect, "power is never considered in terms of the responsibility of choice which is inherent in freedom" since its "only norms are taken from alleged necessity, from either utility or security". But human beings are not completely autonomous. Our freedom fades when it is handed over to the blind forces of the unconscious, of immediate needs, of self-interest, and of violence. In this sense, we stand naked and exposed in the face of our ever-increasing power, lacking the wherewithal to control it. We have certain superficial mechanisms, but we cannot claim to have a sound ethics, a culture and spirituality genuinely capable of setting limits and teaching clear-minded self-restraint.[254]

The Pope's intent here is immediate and salutary. Sometimes we don't want to recognise the limits of our knowledge. "Each age tends to have only a meagre awareness of its own limitations. It is possible that we do not grasp the gravity of the challenges now before us." And our grasp on ethics has become eroded over time

---

[253] *Laudato Si'*, n.102. The citation that the Pope uses here is from John Paul II, *Address to Scientists and Representatives of the United Nations University*, Hiroshima, 25 February 1981, 3.

[254] *Laudato Si'*, n. 105. The two references here are to Romano Guardini, *Das Ende der Neuzeit* 9th ed., Würzburg, 1965, 87-88.

through the excesses in philosophy of utilitarianism, scepticism, nihilism, post-modernism and other "isms" which have so reduced ethics to the hedonic principle (if it feels good it must be right) that our sense of moral responsibility sinks ever more deeply into the quicksands of moral subjectivism. "The risk is growing day by day that man will not use his power as he should"; in effect, "power is never considered in terms of the responsibility of choice which is inherent in freedom" since its "only norms are taken from alleged necessity, from either utility or security".

Some problematic aspects of *Laudato Si'*

However, when it comes to the application of Catholic social teaching to the current problems of the world, this Encyclical seems to adopt a strongly partisan approach both in describing the environmental issues with which human beings are currently dealing and the ways with which they should be dealt. Pope Francis describes his Encyclical as a *"lengthy reflection"*.[255] It would seem, then, not to be simply a teaching document in the traditional sense. Yes, it contains solid Catholic social teaching. But it also contains many statements of "fact" which are disputable, and political and economic opinions which also are disputable. The language used to describe the current ecological condition of the planet is often extreme[256], and unnuanced[257], even apocalyptic[258], in its promoting of preferred conclusions. So, these parts of the Encyclical are a mixed bag and we must, as the Australian Catholic Bishops' Conference rightly reminds us, "consider the content and context of each document, especially the intention of the Pope in issuing

---

255 *Laudato Si'*, n. 246.
256 An example of this is: "The earth, our home, is beginning to look more and more like an immense pile of filth." *Laudato Si'*, n 21.
257 Eg, "Politics and the economy tend to blame each other when it comes to poverty and environmental degradation. It is to be hoped that they can acknowledge their own mistakes and find forms of interaction directed to the common good." *Laudato Si'*, n. 198.
258 *Laudato Si'*, n. 161.

the document, and the way in which the bishops and the whole Church receive the teaching."

In many places, this Encyclical is more akin to a homily or exhortation than to a teaching document. This in itself is not unreasonable since it is written to provoke people to a sense of urgency about the environmental situation as described by the Pope particularly in the earlier parts of the document. But given the pessimistically one-sided approach to the current state of affairs, especially, but not only, where global warming is concerned, it is less than reasonable.

Scientific authorities

It needs to be said that this Encyclical, containing as it does so much good and strong Catholic teaching, nevertheless promotes particular scientific findings which are not sourced. Pope Francis says:

> I will begin by briefly reviewing several aspects of the present ecological crisis, with the aim of drawing on the results of the best scientific research available today, letting them touch us deeply and provide a concrete foundation for the ethical and spiritual itinerary that follows.[259]

He does not say which authorities he considers provides "the best scientific research available" or why he considers these authorities to be better than others. Does he mean the Intergovernmental Panel on Climate Change? He does not say. What he does then conclude is this:

> There is a very solid scientific consensus that now indicates that we are presently witnessing a disturbing warming of the climatic system. In recent decades this warming has been accompanied by a constant rise in the sea level and, it would appear, by an increase in extreme weather events, even if a scientifically determinable cause cannot be assigned to each particular phenomenon. ... It is true that there are other factors (such as volcanic activity,

---

[259] *Laudato Si'*, n. 15.

variations in the Earth's orbit and axis, the solar cycle) yet a number of scientific studies indicate that most global warming in recent decades is due to the great concentration of greenhouse gasses (carbon-dioxide, methane, nitrogen oxides and others) released mainly as a result of human activity. Concentrated in the atmosphere, these gasses do not allow the warmth of the sun's rays reflected by the Earth to be dispersed into space. This problem is aggravated by a model of development based on the intensive use of fossil fuels, which is at the heart of the worldwide energy system. Another determining factor has been the increase in changed uses of the soil, principally deforestation for agricultural purposes.[260]

This is all very well except that it does not confront powerful arguments raised against "consensus science". Here the objection is that "consensus science" is in fact politics not science at all. Professor Ian Plimer is Australia's best known geologist and recognised as a leading historian of climate change. For Ian Plimer, "Science where a majority of votes by climate scientists determines a scientific truth is politics, not science."[261] Moreover, says Plimer[262], "appeals to consensus are not new. The methodology of science allows problems to be solved, whereas the science of the global warmers is designed to confirm a political opinion."[263] For Plimer, the Pope's advisors have not been honest. "The Pope should have been informed by his advisors that the climate 'debate' is not about science. It is about politics."[264]

Paleoclimatologist Professor Robert M Carter is another scholar who has questioned the prevailing "consensus science". In his book *Climate: The Counter Consensus* (2010), and in more measured

---

[260] *Laudato Si'*, n. 23.
[261] Ian Plimer, *Heaven + Earth*, Ballan, Connor Court Publishing, 2009, 450.
[262] Professor Ian Plimer is Australia's best known geologist. He is Emeritus Professor of Earth Science at the University of Melbourne, Victoria. He was Professor on Mining Geology at The University of Adelaide, South Australia (2006-2012).
[263] *Ibid.*, 449.
[264] Ian Plimer, *Heaven + Hell*, Ballarat, Connor Court Publishing, 2015, 38-9.

tones than Plimer, Carter contends that:

> The consensus view of experienced scientists is undoubtedly that both the extreme 'alarmist' and the extreme 'denialist' views of human-caused global warming are incorrect. The majority of scientists, most of whom are independent of the IPCC, espouse views that represent a balanced summary of the available scientific information. Such a view neither inflates nor underestimates the risk of human-caused change, and at the same time takes proper account of the demonstrably severe risks of natural climate change.[265]

Carter, like Plimer, expresses concern about the politicisation of science, and the fact that:

> As a generalization, it can be said that most of the scientific alarm about dangerous climate change is generated by scientists in the meteorological and computer modelling group, whereas many (though not all) geological scientists see no cause for alarm when modern climate change is compared with the climate history that they see every time they stand at an outcrop, or examine a drill core.[266]

And then Tim Ball, writing in 2008, points out that CO_2 emissions, natural and human, amounted to 192-224 Gigatonnes (Gt) of carbon per year. Of this, the "human contribution of 7.5 Gt lies within the error range of the first three natural sources, and the total error range is almost 5 times the human production".[267] In any case, the human contribution is less than 4% of the total.

Or take the contribution of atmospheric physicist, Emeritus Professor Garth Paltridge, in his *The Climate Caper* (2009).

> Perhaps the major characteristic of the science of global warming is the uncertainty associated with just about every aspect of it. Setting aside the difficulties of models and their forecasts that we

---

[265] Robert M Carter, *Climate: The Counter Consensus*, London, Stacey, 2010, 243.
[266] Robert M Carter, *ibid.*, 23.
[267] T Ball, 2008, "Environmentalists seize green moral high ground ignoring science", *Canada Free Press*, 13 June 2008, and cited in Robert M Carter, *ibid.*, 269.

discussed in earlier chapters, the actual measurements of trends in the environment are invariably plagued with inaccuracy. More to the point, in climate research there is no such thing as a controlled experiment – except of course in the overall sense that the earth atmosphere system is already being subjected to a forced change of atmospheric carbon dioxide, and ultimately no doubt we will see what the response of the system turns out to be. Even then it may be that we will not necessarily be able to ascribe the response to the specific cause of CO_2 forcing. A researcher in climate is always faced with the problem, perhaps more familiar to biologists than to physicists, of disentangling a multitude of causes for any particular effect. The study of climate is a statistician's paradise.[268]

Finally, research scientist Roy W Spencer, whose PhD is in meteorology and was a former Senior Scientist for Climate studies at NASA, also rings alarm bells at the current drift of the debate on climate change. In his *Climate Confusion*, Spencer "explains in simple terms how the climate system really works, why man's role in global warming is more myth than science, and how the global warming hype has corrupted Washington and the scientific community".[269] He concludes his book with this observation:

> I often wonder: What motivates politicians and environmentalists who advocate policies that are not only doomed to failure, but also hurt so many other people ... especially the poor?[270]

As I have already remarked earlier, I do not seek to be the arbiter of the scientific debate. But I am aware that despite what some might say this is not a case of "settled science" and each of these authors brings to the table serious scientific literature with which to support their line of argument.

---

268 Garth Paltridge, *The Climate Caper*, Ballan, Connor Court Publishing, 2009, 96-97.
269 Roy W Spencer, *Climate Confusion*, New York, Encounter Books, 2008. From the book sleeve.
270 *Ibid.*, 182.

Other environmental concerns

Prescinding from the debate on "climate change" and whether or not global warming has been largely caused by human beings, there are in the Encyclical some other highly contentious observations about the general state of the environment, again unreferenced. Some of these matters of fact are open to challenge from existing authoritative UN documents, and by some other scholars.

Perhaps the key to understanding the nature of this "lengthy reflection" is to be found in the final paragraph of the opening section of *Laudato Si'* where Pope Francis specifies his intention.

> Although each chapter will have its own subject and specific approach, it will also take up and re-examine important questions previously dealt with. This is particularly the case with a number of themes which will reappear as the Encyclical unfolds. As examples, I will point to the intimate relationship between the poor and the fragility of the planet, the conviction that everything in the world is connected, the critique of new paradigms and forms of power derived from technology, the call to seek other ways of understanding the economy and progress, the value proper to each creature, the human meaning of ecology, **the need for forthright and honest debate**, the serious responsibility of international and local policy, the throwaway culture and the proposal of a new lifestyle. **These questions will not be dealt with once and for all, but reframed and enriched again and again**.[271] [Emphasis added]

Having looked at what constitutes Catholic social teaching, well referred to in this Encyclical, I now want to turn my attention to the highly contentious aspects of the Encyclical and in the spirit of **"the forthright and honest debate"** for which the Pope calls. I will examine some of the factual claims made in *Laudato Si'* and compare them with the factual claims made in internationally respected sources. These sources are:

    1. United Nations, *The Millennium Development Goals Report*

---
271 *Laudato Si'*, n. 16.

2014; and

2. United Nations Department of Economic and Social Affairs Population Division, *World Population Prospects: The 2015 Revision*.

In making these critical comments I am, of course, in no way dissenting from Catholic social teaching, the moral principles which should guide the way society (at both a national and an international level) makes its social and economic arrangements. What is at issue here are the "facts" upon which the Encyclical relies at certain key points, and the way in which those alleged facts are then handled.

Bias in *Laudato Si'*?

Before going on to look at disputed questions of fact, it is important to look at some of the less pleasant comments in the Encyclical directed at those who may well disagree with the description of our current situation favoured by the Encyclical. For example, when it comes to specific environmental issues, Pope Francis appears to be less than objective. In the context of his introductory remarks and when he uses the language of "ecological crisis" he says this:

> Obstructionist attitudes, even on the part of believers, can range from denial of the problem to indifference, nonchalant resignation or blind confidence in technical solutions.[272]

Opposition to the "consensus position" on climate change, for example, may be the position which people have adopted having given the matter serious consideration. Their attitudes are not "obstructionist". Nor should the words "denial" or "denier" be used of those who have another opinion because it crudely calls to mind the use of the word in the context of "holocaust deniers". In any case, what of those who support the "consensus science" because it secures their positions in universities, provides them with grants, and allows them to be seen to be on the "winning side"? If one is

---

272 *Laudato Si'*, n. 14.

going to make a personal criticism of the motives of those with whom you disagree, they may well want to suggest unworthy motives on the part of others. Later in the Encyclical, Pope Francis returns to this theme:

> It must be said that some committed and prayerful Christians, with the excuse of realism and pragmatism, tend to ridicule expressions of concern for the environment. Others are passive; they choose not to change their habits and thus become inconsistent.[273]

To whom is he referring? How many are "some"? Who are the "others"? He identifies no one and no collection of Christians. To identify scepticism of "global warming catastrophe" which has been caused by human activity and whose future catastrophic trajectory can be controlled by human activity, with a tendency to "ridicule expressions of concern for the environment" is grossly unfair to those who hold to such scepticism. People who are very concerned in general with protecting the environment, and do all in their power to mitigate environmental degradation, may not necessarily agree with the theory of human caused catastrophic global warming. This caricature of those who hold to another point of view has allowed former Australian Prime Minister Kevin Rudd to surmise whether the person really being attacked in the Encyclical is Cardinal Pell. Citing this paragraph where the Pope goes on to say that what they all need is an "ecological conversion", Rudd said: "Perhaps the Pope had Cardinal Pell in mind when this paragraph was written for the Encyclical".[274,275]

And the use of the term "ecological conversion" is also misplaced. It first appears early in the Encyclical when Saint John Paul II is cited.

> Saint John Paul II became increasingly concerned about this issue.

---

273 *Laudato Si'*, n. 27.
274 Kevin Rudd, "The 2015 Rowan Williams Lecture", 10th November 2015.
275 Latika Bourke, "Kevin Rudd blasts George Pell over climate change", *Sydney Morning Herald*, 10th November 2015.

> In his first Encyclical he warned that human beings frequently seem "to see no other meaning in their natural environment than what serves for immediate use and consumption". *Subsequently, he would call for a global ecological conversion.*²⁷⁶ [Emphasis added]

The reference to the alleged "subsequent call" was from a *Catechesis* which Saint John Paul II gave two years later on the 17ᵗʰ January 2001. What Saint John Paul II in fact said was this:

> We must therefore encourage and support the "ecological conversion" which in recent decades has made humanity more sensitive to the catastrophe to which it has been heading.²⁷⁷

Saint John Paul II did not "call for an ecological conversion"; he acknowledged an already existing conversion which had been apparent in recent decades. Pope Francis' suggestion that "the committed and prayerful Christians" who don't agree with his take on the science of global warming need to experience "an ecological conversion" is a bit much when many of those who disagree with the politics of global warming are very much committed to a proper treatment by human beings of the environment. The truth is that there has been, as Saint John Paul II put it, a global ecological conversion, and I would say an ecological conversion the likes of which we have never seen before and it should be acknowledged as such. And it is also true that great strides have been made where issues of global poverty, water quality, species protection and the like are concerned. Yes, we have a way to go. Yes, we can do better. But yes, the human family has gone a long way towards redressing those problems.

---

276 *Laudato Si'*, n. 5.
277 John Paul II, *General Audience*, 17 January 2001, http://w2.vatican.va/content/john-paul-ii/en/audiences/2001/documents/hf_jp-ii_aud_20010117.html

The Church and scientific controversies

George Weigel has observed that there is in this Encyclical much more about us human beings, our flaws and our duties to the world in which we live, than, say, climate change. There is a lot that our good *Pastor pastorum* has to say to us:

> ...if you read *Laudato Si'* as an encyclical primarily about us, and not primarily about trees, plankton, and the Tennessee snail darter. Which is, of course, precisely as it should be. For Pope Francis is first and foremost a pastor who, like his predecessor Saint John Paul II, wants to lift up for our reflection what the Polish pontiff called "the nobility of the human vocation to participate responsibly in God's creative plan".

Moreover, Weigel points out that:

> reading *Laudato Si'* as if it were a climate-change encyclical, period, is somewhat akin to reading Moby Dick as if it were a treatise on the 19th-century New England whaling industry. The ships and the harpoons are an important part of the story, to be sure; but if they become the whole story, you miss what Melville's sprawling novel is really about. Ditto with *Laudato Si'*: If you read it as "the global-warming encyclical," you will miss the heart and soul of what this sprawling encyclical is about – which is us. [278]

But what Weigel says here is neither a fair nor accurate account of the Encyclical considered as a whole. It is a relatively small part of the Encyclical that deals with the principles and particulars of Catholic social teaching. But the Encyclical has a lot to say about ecology, human ecology, and climate change. That being the case it would be good if the facts in it were correct, and where they are contentious due deference be given to alternative points of view. The Church is not there to resolve scientific and related kinds of

---

278 Cf http://www.nationalreview.com/article/419933/parsing-popes-ecology-encyclical-its-about-lot-more-climate-change-george-weigel

controversial opinion. Indeed, Pope Francis himself says this:

> There are certain environmental issues where it is not easy to achieve a broad consensus. Here I would state once more that *the Church does not presume to settle scientific questions or to replace politics*. But I am concerned to encourage an honest and open debate so that particular interests or ideologies will not prejudice the common good.[279] [Emphasis added]

Be that as it may, it does not assist "honest and open debate" when the scientific data from one point of view is privileged above all other points view to the degree that the alternative views are neither identified nor discussed. The science of "climate change" is not settled and the debate continues.[280] And nor does it assist informed debate when what the Encyclical says about water, biodiversity and the lot of the poor is in such obvious conflict with the data produced in reports by the UN without any explanation as to why those reports have not been factored into consideration. Indeed, this data is neither referred to nor in any other way acknowledged.

Pope Francis and deserts

In paragraph 217 the Pope cites the authority of his predecessor, Pope Benedict XVI: "The external deserts in the world are growing, because the internal deserts have become so vast". Here both Popes are referring to the fact that the interior lives of people have been laid waste through moral corruption, a corruption which is linked to the increased desertification of the planet as we go about seeking our own convenience at the expense of the planet. But the remark about the increasing desertification of our planet is clearly meant

---

279 *Laudato Si'*, n. 188.
280 Cf, for example, Ian Plimer's four books, and the works of Robert M Carter, Garth Paltridge, and Roy Spencer referred to above, and Miranda Devine, "Perth electrical engineer's discovery will change climate change debate", 3rd October 2015, http://www.news.com.au/national/western-australia/miranda-devine-perth-electrical-engineers-discovery-will-change-climate-change-debate/story-fnj4anv2-1227555674611

to be a factual one. However, Pope Benedict made this comment on the 24th April 2005, in a homily which is not, of course, at the level of an encyclical where authority is concerned. And in 2005 the view that the deserts of the world were expanding was the then received account of the facts as best they could then be known. Lester Brown from the Earth Policy Institute described the state of affairs as it was believed to obtain at the time in these terms:

> Our early twenty-first century civilization is being squeezed between advancing deserts and rising seas. Measured by the land area that can support human habitation, the earth is shrinking. Mounting population densities, once generated solely by the addition of over 70 million people per year, are now also fuelled by the relentless advance of deserts and the rise in sea level.
>
> The newly established trends of expanding deserts and rising seas are both of human origin. The former is primarily the result of overstocking grasslands and overplowing (*sic*) land. Rising seas result from temperature increases set in motion by carbon released from the burning of fossil fuels.[281]

But by 2009, the understanding of what was happening was being greatly revised.

> It had been expected that as a result of global warming the world's deserts would increase in size. Now comes a new prediction from a group of scientists that global warming is helping shrink deserts, primarily the Sahara. New satellite images are showing that the Sahara is not growing, but new patches of green growth areas are being developed. All due to global warming according to the scientists. Global warming is increasing the amount of rainfall the world is getting now and is helping turn the tide of desert growth. The growth and shrinking of deserts is a little bit more complex than getting a higher yield of rainfall. The over use of lands for agriculture hasten the creation of new deserts.

---

281 Lester R. Brown, "The Earth Is Shrinking: Advancing Deserts and Rising Seas Squeezing Civilization", 15 November 2006, http://www.earth-policy.org/plan_b_updates/2006/update61

> Areas in Egypt and even in Italy have seen increases in once tillable lands turning into deserts. Ground contamination from pesticides and fertilizers has impacted ground aquifers that are used in agriculture and has created brackish water that cannot support farming. As this process occurs lands begin to turn into deserts. The impact from global warming has not been fully played out yet, and there are scientific arguments from both sides of the aisle disputing their respective positions on this matter. But, if deserts that were not used in farming can be reclaimed, then this is an unexpected positive from global warming.[282]

And this from the *New Scientist*:

> Burkina Faso, one of the West African countries devastated by drought and advancing deserts 20 years ago, is growing so much greener that families who fled to wetter coastal regions are starting to go home.
> New research confirming this remarkable environmental turnaround is to be presented to Burkina Faso's ministers and international aid agencies in November. And it is not just Burkina Faso.
> New Scientist has learned that a separate analysis of satellite images completed this summer reveals that dunes are retreating right across the Sahel region on the southern edge of the Sahara desert. Vegetation is ousting sand across a swathe of land stretching from Mauritania on the shores of the Atlantic to Eritrea 6000 kilometres away on the Red Sea coast.[283]

And in his important scholarly work on global warming, Professor Ian Plimer sets out the issue of desertification very carefully.

> To blithely state that global warming will lead to desertification ignores previous validated science that shows that desertification occurs during glaciation ... Modern climate change can lead to widespread and irreversible desertification, or so we are told ... Evidence is now coming to light that shows that Africa is not

---

[282] By John Vlahakis, "Are deserts shrinking?" 20 July 2009, http://www.earthyreport.com/site/are-deserts-shrinking/

[283] "Africa's deserts are in "spectacular" retreat", *New Scientist*, 18 September 2002.

undergoing severe desertification. In fact recent evidence shows the opposite.[284]

It is not for me to resolve the question as to what is a desert (is Antarctica a desert?) or the science involved in measuring these things. But this debate has been well known for years, and very soon after Pope Benedict's observation. Why did someone not advise the current Holy Father that the situation had changed, and that there was a whole new debate going on involving global warming and its effects?

Provocative language and questionable facts

Very early on, (in fact in n. 2 of *Laudato Si'*), we are alerted to the propensity for extreme and provocative language:

> This sister [earth] now cries out to us because of the harm we have inflicted on her by our irresponsible use and abuse of the goods with which God has endowed her. We have come to see ourselves as her lords and masters, entitled to plunder her at will. The violence present in our hearts, wounded by sin, is also reflected in the symptoms of sickness evident in the soil, in the water, in the air and in all forms of life. This is why the earth herself, burdened and laid waste, is among the most abandoned and maltreated of our poor; she "groans in travail" (Rom 8:22). We have forgotten that we ourselves are dust of the earth (cf. Gen 2:7); our very bodies are made up of her elements, we breathe her air and we receive life and refreshment from her waters.

Immediately we are confronted by a mixture of sweeping claims, couched in extremely provocative language, the factual basis of which we would expect to see justified somewhere in the text, and fair comment. For example:

1. The earth is crying out to us because of human abuse;

---

[284] Ian Plimer, *Heaven + Earth: Global Warming: The Missing Science*, Ballan, Victoria, Connor Court Publishing, 2009, 204.

2. We see ourselves as lords and masters who "*plunder*[285] her (earth) at will"; [emphasis added]
3. Human beings are depicted as having violence present in their hearts[286], and the sin of man is reflected in the symptoms of sickness we find in soil, water, air, and in all forms of life; [This is a very unnuanced account of the nature of man]; and
4. We have forgotten that we are made of dust of the earth and our relationship to the created order which is needed to sustain life.

Pope Francis states that the "earth herself, burdened and laid waste, is among the most abandoned and maltreated of our poor; she 'groans in travail'". Here we are referred to the *Letter of St Paul to the Romans* 8:22 as justification for the statement. This is, at best, a very forced use of the Biblical text. That text is referring to the consequences of the sin of Adam which will only be resolved or alleviated at the end of time, when the children of God are revealed and the whole material creation is renewed. It does not refer to environmental politics and policies.[287]

The Holy Father goes on to cite the teachings of previous Popes, teachings which are far more restrained than his own in their appeal to facts and in their use of language.

An example of this may be found in paragraph 6 of the Encyclical where Francis begins by citing Pope Benedict XVI from an address he gave in 1987. We should note that the then Cardinal Ratzinger

---

285 "Plunder" means: "forcibly stealing goods from, especially, in time of war or civil disorder." Concise Oxford English Dictionary 2011.
286 Cf CCC: "Although it is proper to each individual, original sin does not have the character of a personal fault in any of Adam's descendants. It is a deprivation of original holiness and justice, but human nature has not been totally corrupted: it is wounded in the natural powers proper to it, subject to ignorance, suffering and the dominion of death, and inclined to sin - an inclination to evil that is called concupiscence". Baptism, by imparting the life of Christ's grace, erases original sin and turns a man back towards God, but the consequences for nature, weakened and inclined to evil, persist in man and summon him to spiritual battle.
287 Cf *Catechism of the Catholic Church*, n. 2630; Joseph A Fitzmyer SJ, "The Letter to the Romans", in Raymond E Brown, SS, Joseph A Fitzmyer, SJ, and Roland E Murphy, O.Carm, *The New Jerome Biblical Commentary*, London, Geoffrey Chapman, 1989, 854.

was speaking in 1987, 28 years ago. Since then there have been great improvements to the lot of the poor. [See the section on water below.]

> Among the key issues, how can we not think of the millions of people, especially women and children, who lack water, food, or shelter? The *worsening scandal* of hunger is unacceptable in a world which has the resources, the knowledge, and the means available to bring it to an end. It impels us to change our way of life, *it reminds us of the urgent need to eliminate the structural causes of global economic dysfunction and to correct models of growth that seem incapable of guaranteeing respect for the environment and for integral human development, both now and in the future.* Once again I invite the leaders of the wealthiest nations to take the necessary steps to ensure that poor countries, which often have a wealth of natural resources, are able to benefit from the fruits of goods that are rightfully theirs.[288] [Emphasis added]

He then refers to the following quotes from Pope Benedict's 2009 Encyclical letter *Caritas in veritate*:

> The deterioration of nature is in fact closely connected to the culture that shapes human coexistence.[289]
>
> The book of nature is one and indivisible: it takes in not only the environment but also life, sexuality, marriage, the family, social relations.[290]

Pope Francis summarises Pope Benedict as saying that "the natural environment has been *gravely damaged by our irresponsible behaviour*". [Emphasis added] Pope Benedict did not use such extreme language. On the contrary, the whole tone of Benedict is restrained and appropriate. But this use of extreme language ("worsening scandal", "gravely damaged by our irresponsible behaviour") is characteristic of the Encyclical whenever the

---

288 Address of His Holiness Pope Benedict XVI to The Diplomatic Corps Accredited To The Holy See For The Traditional Exchange of New Year Greetings, Monday, 8 January 2007.
289 *Caritas in Veritate*, 29 June 2009, n. 51 second paragraph.
290 *Ibid*, third paragraph.

discussion moves to environmental degradation. In another place, Francis claims that "young people demand change".²⁹¹ Do they? Which young people and in which countries? Nowhere does he justify this proposition. This may be because it is simply not possible to do so.

"Fact" and who says so?

Chapter one of *Laudato Si'* is long on assertion about scientific and related matters but short on referenced fact. The difficulty here is that climate science and related specialities in botany, geology, astronomy, and biology are highly specialised areas. So we need to know who is really speaking to us through this Encyclical, who the authorities are upon which the Pope relies. But we are simply not told. When he makes this quote, with approval, "Who turned the wonderworld of the seas into the underwater cemeteries bereft of colour and life?"²⁹² his referenced authority is the Catholic Bishops' Conference of the Philippines, which is hardly a scientific source.

Water

The Encyclical makes this claim:

> "One particularly serious problem is the quality of water available to the poor ... *the quality of available water is constantly diminishing ...*"²⁹³ [Emphasis added]

Now compare that statement with the UN's *The Millennium Development Goals 2014 Report*. It says that over the past twenty years:

> the target of halving the proportion of people without access to an improved drinking water source was achieved in 2010, five years ahead of schedule. In 2012, 89 percent of the world's population had access to an improved source, up from 76 per cent in 1990.

---

291 *Laudato Si'*, n. 13.
292 *Laudato Si'*, n. 41 and cf footnote reference 25.
293 *Laudato Si'*, nn 29 & 30.

Over 2.3 billion people gained access to an improved source of drinking water between 1990 and 2012.[294]

So it would seem that the *"quality of available water"* is not only not diminishing, the situation is greatly improving. If the UN has got it wrong, then what is the authority to show that the UN has got it wrong?

Biodiversity

The Encyclical addresses the matter of biodiversity.[295] Pope Francis says:

> Each year sees the disappearance of thousands of plant and animal species which we will never know, which our children will never see, because they have been lost for ever. *The great majority become extinct for reasons related to human activity.* Because of us, thousands of species will no longer give glory to God by their very existence, nor convey their message to us. We have no such right."[296] [Emphasis added]

The first thing to notice is that we have been losing species to extinction since time immemorial. There have been five previous mass extinctions of species during the earth's history, the last being the extinction of the dinosaurs.[297] Human activity can hardly be blamed for any of those mass extinctions! Today, the situation is different with the impact of human activity being implicated in the current episode of mass extinction of species.[298] The point I make here is that mass extinctions are not new. Nor is it clear that all current species extinctions and reductions in species populations are caused by human beings. Nevertheless, as Pope Francis says, we need to understand and accept our responsibility as human beings

---

294 United Nations, *The Millennium Development Goals Report 2014*, p. 4.
295 *Laudato Si'*, n. 32-42.
296 *Laudato Si'*, n.33.
297 http://www.biologicaldiversity.org/programs/biodiversity/elements_of_biodiversity/extinction_crisis/
298 ibid.

to minimise negative impacts of human activity to slow down the process of the extinctions of other species, and to preserve the natural habitat in so far as we can. And this is, in fact, what we are doing.

Second, The Millennium Development Goals 2014 Report says this about these contemporary mass extinctions:

> The Red List Index shows that Biological diversity provides many different ecosystem services upon which human lives and livelihoods depend. For example, many studies have shown that declines or absences of species that pollinate crops lead to reduced crop productivity and value. A recent analysis of the Red List Index revealed declining population and distribution trends and increasing extinction risk of pollinator bird and mammal species—a result likely to be mirrored by insect pollinators. *More needs to be done to reverse these trends, reduce extinction rates and, hence, safeguard the benefits species provide to society.*[299] [Emphasis added]

This statement is nowhere near the Pope's claim: "Each year sees the disappearance of thousands of plant and animal species which we will never know, which our children will never see, because they have been lost for ever. The great majority become extinct for reasons related to human activity." Yes, the UN Report says that "overall, species are declining in population and distribution and, hence, moving faster towards extinction", but nowhere does it speak of the "***disappearance*** of thousands of ... species" each year. A "declining population" is not the same as "disappearance" of a population. In so far as we can do something about cases where human activity is not the cause of a declining population, that is what we are doing.

It also seems to have been established that many species have seen a reduction in their natural habitats due to human activity such

---

299 United Nations, *The Millennium Development Goals Report 2014*, p. 43.

as deforestation with an associated reduction in their populations.[300] This is why the United Nations in 1993 passed its *Convention on Biological Diversity*. Pursuant to that *Convention*, achievement goals were adopted to set aside 17 per cent of global terrestrial areas and 10 per cent of coastal and marine areas by 2020 as nature preserves.[301] By 2014 these ecological preserves included some "14.6 percent of the earth's land surface area and 9.7 per cent of its coastal marine areas".[302]

So, it would seem that we have got very close to achieving these internationally agreed upon goals of protecting biodiversity and reducing anthropogenic caused species extinction ahead of time. Yet the Encyclical, while conceding that some places have done some things to offset challenges to biodiversity, speaks as if these issues have been insufficiently addressed. The Encyclical again uses the language of "plunder" under the heading of "Loss of Biodiversity" in its introductory sentence:

> The earth's resources are also being plundered because of short-sighted approaches to the economy, commerce and production. The loss of forests and woodlands entails the loss of species which may constitute extremely important resources in the future, not only for food but also for curing disease and other uses.[303]

The first sentence sets the scene of doom and gloom and the situation is hardly retrieved by later admissions that "some countries have made significant progress in establishing sanctuaries on land and in

---

300 Cf the report of the US National Wildlife Federation, https://www.nwf.org/Wildlife/Threats-to-Wildlife/Habitat-Loss.aspx

301 "Target 11: By 2020, at least 17 per cent of terrestrial and inland water, and 10 per cent of coastal and marine areas, especially areas of particular importance for biodiversity and ecosystem services, are conserved through effectively and equitably managed, ecologically representative and well connected systems of protected areas and other effective area-based conservation measures, and integrated into the wider landscapes and seascapes." See Aichi Biodiversity Targets under the UN's *Convention on Biological Diversity* which entered into force on the 29th December 1993.

302 United Nations, *The Millennium Development Goals Report 2014*, p. 43.

303 *Laudato Si'*, n. 32.

the oceans ..."³⁰⁴ Moreover, sweeping statements about forests need qualification. Here is not the time or place to go into this in great detail, but the following points need to be taken into account.

- In the period 2000-2005, South America reported the largest net loss of forest, followed by Africa;
- In the period 2000- 2005 Asia showed a net gain of forests of around 1 million hectares per year, despite high rates of deforestation in many countries in the region, in particular in Southeast Asia;
- This net gain is attributed to large-scale afforestation [convert land into forest for commercial use], particularly in China, where there has been an annual increase of more than 4 million hectares;
- Meanwhile in Europe forest areas continued to expand, although at a relatively slow rate, while North and Central America and Oceania registered a relatively small annual net loss of forests over the 1990-2005 period;
- The five countries with the largest annual net loss of forest area in the period 2000-2005 were Brazil, Indonesia, Sudan, Myanmar and Zambia;
- The five countries with the largest annual net gain in forest area over the same period were China, Spain, Vietnam, the United States and Italy; and
- Chile, Costa Rica, India and Vietnam are among the countries which have recently recorded a change from having a net loss of forests to having a net gain in forest area.³⁰⁵

Over time land has been cleared for a number of purposes, including making land available for agriculture precisely to grow more food for more people. It is difficult to argue against clearing land to grow more food. What is called for is greater efficiency of land use for growing food and afforestation wherever possible. In

---

304 *Laudato Si'*, n. 37.
305 All of this information and more is readily available at the website of the *United Nations Environment Programme*, http://www.unep.org/vitalforest/Report/VFG-02-Forest-losses-and-gains.pdf

countries like those to be found on the African continent, absent access to cheap forms of heating, people cannot be blamed when they chop down trees to use as fuel for cooking and for keeping warm.

The following facts need to be kept in mind:

1. In developing countries, especially in rural areas, 2.5 billion people rely on biomass, such as fuelwood, charcoal, agricultural waste and animal dung, to meet their energy needs for cooking.
2. In the absence of new policies, the number of people relying on biomass will increase to over 2.6 billion by 2015 and to 2.7 billion by 2030 because of population growth.
3. About 1.3 million people – mostly women and children – die prematurely every year because of exposure to indoor air pollution from biomass. Valuable time and effort is devoted to fuel collection instead of education or income generation. Environmental damage can also result, such as land degradation and regional air pollution.
4. Halving the number of households using traditional biomass for cooking by 2015 – a recommendation of the United Nations Millennium Project – would involve 1.3 billion people switching to other fuels.
5. There is evidence that, in areas where local prices have adjusted to recent high international energy prices, the shift to cleaner, more efficient use of energy for cooking has actually slowed and even reversed.[306]

The need to replace what has been removed has been recognised for a long time now and significant efforts at afforestation have been successful, although much more needs to be done. Perhaps the proximity of Argentina to Brazil has made the Argentinian Pope Francis particularly alert to this significant ecological issue. But a balance in assessment has to be kept, acknowledging where

---

306 *Energy for Cooking in Developing Countries*, https://www.iea.org/publications/freepublications/publication/cooking.pdf

improvement in the situation has occurred while also maintaining the need for the world to do more. But this is not an issue to which the world has not yet responded.

The quality of life

The Encyclical says:

> ... we cannot fail to consider the effects on people's lives of environmental deterioration... [307]
> ... the growth of the past two centuries has not always led to ... an improvement in the quality of life.[308]

But looking at the data, what do we see?

1815 there were approximately 1 billion people alive on the planet[309], the average lifespan was 30 years[310], and the per capita income was a mere $100. The life of man in nature, as Thomas Hobbes famously remarked, was nasty, poor, brutish and short.

2015 there are 7.2 billion people living on Planet Earth[311], the average lifespan is 71[312], and the GDP per capita, using purchasing power parity, is over $12,000. In other words, as our numbers have grown so has our well-being and prosperity.[313]

Contrary to what the Encyclical states, there has been an unprecedented improvement in the quality of life of human beings globally with respect to the key performance indicators of lifespan

---

307 *Laudato Si'*, n. 43.
308 *Laudato Si'*, n. 46.
309 http://www.worldometers.info/world-population/
310 Cf for example http://www.conferenceboard.ca/hcp/details/health/life-expectancy.aspx
311 http://wdi.worldbank.org/table/2.1
312 World Health Organisation on 15th May 2014: "People everywhere are living longer, according to the "World Health Statistics 2014" published today by WHO. Based on global averages, a girl who was born in 2012 can expect to live to around 73 years, and a boy to the age of 68. This is six years longer than the average global life expectancy for a child born in 1990." http://www.who.int/mediacentre/news/releases/2014/world-health-statistics-2014/en/
313 Steve Mosher, http://www.lifeissues.net/writers/mos/mos_338earthnotsick.html

and prosperity. And it may well be that at least declines in population of some species, and extinctions of other species, are a by-product of these monumental achievements.

But quality of life concerns far more than longevity. It also involves human happiness. The collapse of moral codes in relation to human sexuality and respect for life has brought misery and unhappiness, particularly but not exclusively in the West. Here the only work being done by governments is to increase the amount of human unhappiness by subverting cultures which have a better grasp of the natural law, a better grasp of what it means to speak of the inherent dignity of the human being from conception until natural death. In this sense those governments act against the integral ecology to which the Pope refers, undermining the stability of the family which is the fundamental group unit of society.

Global Inequality

The encyclical:

> It needs to be said that, generally speaking, there is little in the way of clear awareness of problems which especially affect the excluded. Yet they are the majority of the planet's population, billions of people. These days, they are mentioned in international political and economic discussions, but one often has the impression that their problems are brought up as an afterthought, a question which gets added almost out of duty or in a tangential way, if not treated merely as collateral damage. Indeed, when all is said and done, they frequently remain at the bottom of the pile.[314]

In other words, it is alleged that the world as a whole is showing no interest in the poor and excluded. They are at best something to be talked about as "an afterthought", but in reality nobody much cares. They are merely "collateral damage" as the rich continue on their merry way to advance their own projects of self-interest.

---

314 *Laudato Si'*, n. 49.

This seems to me to be not only untrue but also grossly unfair to all those agencies both local and international, supported by donations from churches, other religious foundations, governments, businesses and individuals.

As can be seen in the table below most of the top donor states where foreign aid is concerned, are countries which are democratic and which are based upon the Western Christian tradition. A notable exception to this is the United Arab Emirates which gives 1.25% of its gross national income to foreign aid, more than any other nation. Of course we all can and should do more. But sweeping generalisations of the kind to be found in this Encyclical would not appear to be entirely fair.

On the other hand, it needs to be said, that the same countries which give billions of dollars in international aid also give billions of dollars to anti-human-life bodies such as International Planned Parenthood, Marie Stopes, and the United Nations Population Fund (UNFPA[315]). No reference is made in this Encyclical to the funding of the obvious violation of the human rights of unborn babies in which these organisations are so deeply complicit. Worse still, foreign aid is often tied to poor recipient countries accepting the noxious practices of abortion, contraception, and the homosexual agenda. Thus is violence done to cultures which have a far better appreciation of marriage, family, and the protection of children.

Development assistance by country as percentage of Gross National Income 2013

The Organization for Economic Co-operation and Development also lists countries by the amount of ODA they give *as a percentage of their gross national income*. Five countries met the longstanding UN target for an ODA/GNI ratio of 0.7% in 2013:

Norway – 1.07%

---

[315] The United Nations Population Fund (UNFPA) was formerly known as the United Nations Fund for Population Activities, hence the continuing use of the acronym UNFPA.

Sweden – 1.02%
Luxembourg – 1.00%
Denmark – 0.85%
United Kingdom – 0.72%
Netherlands – 0.67%
Finland – 0.55%
Switzerland – 0.47%
Belgium – 0.45%
Ireland – 0.45%
France – 0.41%
Germany – 0.38%
Australia – 0.34%
Austria – 0.28%
Canada – 0.27%
New Zealand – 0.26%
Iceland – 0.26%
Japan – 0.23%
Portugal – 0.23%
United States – 0.19%
Spain – 0.16%
Italy – 0.16%
South Korea – 0.13%
Slovenia – 0.13%
Greece – 0.13%
Czech Republic – 0.11%
Poland – 0.10%
Slovak Republic – 0.09%

European Union countries that are members of the Development Assistance Committee gave 0.42% of GNI (excluding the $15.93 billion given by EU Institutions).

Non-DAC countries reported preliminary ODA figures as a percentage of GNI as follows:
United Arab Emirates – 1.25%
Turkey – 0.42%
Estonia – 0.13%
Hungary – 0.10%

Latvia – 0.08%
Israel – 0.07%
Russia – 0.03%

The UAE's ODA/GNI ratio rose to 1.25% in 2013, the largest reported share of any country, due to their contributions to address financial and infrastructure needs in Egypt,[1] Jordan, Palestine, Afghanistan and Syria. The UAE puts this figure at 1.33% of GNI and a total of $5.9 in aid contributions for 2013.[316]

In a recently published paper[317] I drew attention to the *World Population Prospects: The 2015 Revision*, prepared by the United Nations' Department of Economic and Social Affairs Population Division[318] and to an excellent analysis of that report by Jonathan Abbamonte.[319] This UN document demonstrates the way in which the world has been seriously misled by population controllers, including international bodies such as UNFPA and USAID, who have been encouraging and even coercing developing countries to embrace their population control policies, policies which have as their centrepiece sterilisation and abortifacient contraception pills and devices coupled with bribes and other methods of coercion.

The latest figures predict that world population will rise from 7.3 billion today to 9.7 billion by 2050 and 11.2 billion by 2100. The default position of population controllers has always been that population growth has been and will be the single biggest cause of catastrophic outcomes for the planet:

1. widespread famines as the world struggles to feed the people;
2. continuing low life expectancy in developing countries; and
3. high infant mortality rates among developing countries.

But, as Abbamonte points out, relying on the information contained in the new Report, the facts are actually these:

---

316 https://en.wikipedia.org/wiki/List_of_governments_by_development_aid
317 http://spuc-director.blogspot.com.au/search?updated-max=2015-08-21T14:02:00%2B01:00
318 http://esa.un.org/unpd/wpp/publications/files/key_findings_wpp_2015.pdf
319 https://www.pop.org/content/no-need-population-control

1. Despite the rapid rise in population over the last two and a half decades, "the percentage of people living with hunger in developing countries has actually dropped from 24% to 14% over the same time period."
2. "World average life expectancy at birth in the early 1950's was 48 years for women and 45 for men. Today those numbers are 73 for women and 68 for men. By 2100, life expectancy at birth will have risen to almost 85 for women and 82 for men worldwide and even higher in developed nations – 92 years for women."
3. Infant and childhood mortality are set to decline sharply worldwide. "By 2100, the rate of deaths among children under the age of five will fall as much as 82% in less developed nations and 80% in the world's least developed countries."

The same case has been well summarised in *The Millennium Development Goals Report 2014*.

> The world has made remarkable progress in reducing extreme poverty. In 1990, close to half of the people in developing regions lived on less than $1.25 a day. This rate dropped to 22 per cent by 2010. This means that the world reached the MDG target – of halving the proportion of people living in extreme poverty – five years ahead of the 2015 deadline. Meantime, the absolute number of people living in extreme poverty fell from 1.9 billion in 1990 to 1.2 billion in 2010. Despite this overall achievement, progress on poverty reduction has been uneven. Some regions, such as Eastern Asia and SouthEastern Asia, have met the target of halving the extreme poverty rate, whereas other regions, such as sub-Saharan Africa and Southern Asia, still lag behind. According to World Bank projections, sub-Saharan Africa will be unlikely to meet the target by 2015.[320]

And again:

> A total of 842 million people, or about one in eight people in the world, were estimated to be suffering from chronic hunger in 2011-2013. The vast majority of those people (827 million) resided in developing regions. Since 1990-1992, significant

---
320 *The Millennium Development Goals Report 2014*, p. 9

progress towards the MDG hunger target has been recorded in those regions. The proportion of undernourished people – those individuals not being able to obtain enough food regularly to conduct an active and healthy life – decreased from 23.6 per cent in 1990-1992 to 14.3 per cent in 2011-2013. However, progress during the past decade was slower compared to that recorded in the 1990s. Should the average annual decline of the past 21 years continue on to 2015, the prevalence of undernourishment would barely exceed the target by about 1 percentage point. Meeting the target, therefore, will require considerable – and immediate – additional effort, especially in countries which have showed little headway.[321]

For the purposes of this book it is enough to show that the world has made rapid progress in significantly reducing hunger and poverty, increasing life expectancy, and lowering infant mortality rates. Some countries still lag well behind. But overall the facts seem to be, at least in some significant measure, in conflict with the Encyclical's claims.

High Rhetoric meets Doomsday

The extreme tone of the encyclical where the degradation of the environment is concerned reaches its zenith at n. 161. Redolent of the excessive and inaccurate predictions of the Club of Rome (cf its report *The Limits to Growth*, 1972) and Paul Ehrlich (*The Population Bomb*, 1968[322]) the Encyclical says this:

> *Doomsday predictions can no longer be met with irony or disdain. We may well be leaving to coming generations debris, desolation and filth.* The pace of consumption, waste and environmental change has so stretched the planet's capacity that our contemporary lifestyle, unsustainable as it is, can only precipitate catastrophes, such as those which even now periodically occur in different areas

---

321 *The Millennium Development Goals Report 2014*, p. 12.
322 Ehrlich warned of the mass starvation of humans in the 1970s and 1980s due to overpopulation. He warned that immediate action needed to be taken to limit population growth.

of the world. The effects of the present imbalance can only be reduced by our decisive action, here and now. We need to reflect on our accountability before those who will have to endure the dire consequences.[323] [Emphasis added]

This passage hardly reflects the objectivity and fairness we might expect. Prophecies of doom have a most unfortunate history. They nearly always fail to eventuate, especially when predicted on the global scale envisioned here. Of course all recent popes have reminded us about our responsibility as human beings properly to care for the world in which we live. A healthy natural environment is needed for the natural health and well-being of human beings. But the ways in which the world responds to the challenges of environmental degradation is for governments to decide, for lay people expert in various fields to prescribe. The Church has no special competence to predict catastrophic outcomes in this area where science and public policy decision makers are struggling to find the necessary corrective measures and in a way that doesn't bring with it unforeseen side effects at least as detrimental as the problems we seek to overcome. But the Church has the responsibility to lay down the moral framework within which environmental problems can be addressed.

Climate Change

We come now to consider one of the central themes of this Encyclical which is climate change or man-made global warming. It is clear that the Pope has lined up with the politically dominant group of climate scientists who warn of environmental catastrophe and recommend economically regressive solutions on the basis of their kind of science, while ignoring contradictory inputs from other disciplines who study this subject. Critics of the dominant group such as Professor Ian Plimer, Australia's best known geologist and recognised leading historian of climate change, together with

---

323 *Laudato Si'*, n. 161.

Robert M Carter, Garth Paltridge, and Roy Spencer beg to differ. In the current political climate critics of "consensus science" on climate change sometimes have to shout to be heard:

> My criticism of the climate industry is that much of the data (eg temperature and CO_2 measurements and corrections) is contentious, that computers have to be tortured to obtain the pre-ordained result, that computer codes are not freely available, that neither the data nor the conclusions are in accord with what we know from the present and the past, that publication is within a closed system that challenges the veracity of the peer-review system, that financial interests are not declared, that the financial rewards for publishing a scary scenario are tempting and that the process of refutation seems to have been overlooked.[324]

Plimer's four books on the subject are not to be wished away. The vast amount of reference material he provides which questions climate consensus cannot simply be ignored. Plimer is not alone in being heavily critical of the "facts" and methodology which is the basis of the "consensus science". His work, and that of many others referred to in chapter 5, is notoriously either overlooked or ridiculed by the consensus scientists. Given that the science on climate issues is not settled other points of view should have been taken into account in this Encyclical. Perhaps the Pope's advisers did not give as full an account of the state of play as they might have.

Whatever about that, I do not myself intend to do what a non-scientist should not do. It is not up to me to resolve the science. But nor should I be stampeded into accepting one particular point of view for fear of being labelled a "denier", "an obstructionist", or simply one who is indifferent to whatever happens to the environment. Like most Catholics I am vitally interested in issues around environmental degradation, of the ways in which we can use renewable energy sources which are cheap and efficient without, at the same time, abandoning the use of cheap coal unless and until we can find something as cheap, as abundant, and as efficient. And I am committed

---

324 Ian Plimer, *Heaven + Hell*, Ballarat, Connor Court Publishing, 90-91.

to Catholic social teaching as a loyal son of the Church.

The issue here is not so much concern for the increased price of electricity in developed countries, although that is still a concern for the poor who need to be able to pay for their basic needs such as heating and cooling, cooking, refrigeration and the like. The much bigger issue here, and the one highlighted in Plimer's newest book, is that without coal many parts of the world will find their efforts to alleviate hunger and poverty defeated.

The really important issue here is that, even if the consensus account of climate change science proves correct, and the suggested policies to ameliorate climate warming are implemented, there may be a massive and unprecedented impact on the lifestyles of people everywhere. Nowhere will this be felt more than among those who are already the poorest and most disadvantaged.

In other words, the very issues that concern the Pope elsewhere in *Laudato Si'* (poverty, hunger, quality of life) may take a sharp turn for the worse among those least able to withstand such changes. This is the point made strongly by Ian Plimer in his most recent book. Here he argues that the solutions proposed by Pope Francis to meet the "climate crisis" as he believes it to be, would be to "condemn the poor to eternal poverty."[325]

To some extent though, Pope Francis recognises this problem, and suggests the need for some short term compromise where coal, for example is still needed absent any other competitive alternative.

> We know that technology based on the use of highly polluting fossil fuels – especially coal, but also oil and, to a lesser degree, gas – needs to be progressively replaced without delay. Until greater progress is made in developing widely accessible sources of renewable energy, it is legitimate to choose the lesser of two evils or to find short-term solutions.[326]

---

325 Ian Plimer, *Heaven + Hell*, Ballarat, Connor Court Publishing. The subtitle of the book: "The Pope condemns the poor to eternal poverty."
326 *Laudato Si'*, n. 165.

I remember very well how, in the 1970s, we were told that we would all freeze to death. Now we are told that the opposite is in fact true, that we are in for inhumanly hot conditions in which we will cook to death. In fairness it needs to be said that there was no evidence of a "consensus" among climate scientists about global cooling in the 1970s. But nevertheless the scare was widely announced in the mass media. For example, a 1974 Times Magazine article, "Another Ice Age?" painted a particularly bleak picture:

> When meteorologists take an average of temperatures around the globe, they find that the atmosphere has been growing gradually cooler for the past three decades. The trend shows no indication of reversing. Climatological Cassandras are becoming increasingly apprehensive, for the weather aberrations they are studying may be the harbinger of another ice age.[327]

But the use of scare tactics goes on. Ian Plimer reminds us with this citation from Stephen Schneider about how truth and attention-seeking do not always make happy bed-fellows. Stephen Henry Schneider (1945-2010) was Professor of Environmental Biology and Global Change at Stanford University, a Co-Director at the Centre for Environment Science and Policy of the Freeman Spogli Institute for International Studies and Senior Fellow in the Stanford Woods Institute for the Environment. His research included modelling of the atmosphere, climate change, and the effect of global climate change on biological systems.[328]

> We need to get some broad-based support to capture the public's imagination. That, of course, entails getting loads of media coverage. So we have to offer up scary scenarios, make simplified, dramatic statements, and make little mention of any doubts we may have ... Each of us has to decide the right balance between being effective and being honest.[329]

---

327 http://www.skepticalscience.com/What-1970s-science-said-about-global-cooling.html
328 These details have been taken from https://en.wikipedia.org/wiki/Stephen_Schneider
329 Stephen Schneider, October 1989 interview for *Discover* magazine and cited in Ian Plimer, *Heaven + Hell*, Ballarat, Connor Court Publishing, footnote 152 on p. 91.

To give some impression of the flavour of the current debate from an alternative point of view, Ian Plimer argues that a number of claims made by those whom he calls "climate catastrophists" have proved to be wrong. Among those he contends that have turned out to be wrong are these:

- Unprecedented temperature changes over the last 100 years;
- Warmest year on record was 2014;
- The ice sheets and glaciers are melting and sea ice is contracting;
- Sea level is rising rapidly, Pacific islands, the Maldives and coastal cities will be inundated;
- Rainfall extreme weather events have increased;
- Global warming will bring more drought;
- Increasing human emissions of CO_2 are affecting global temperature; and
- Sophisticated computer models can be used to predict future climate.[330]

Plimer produces an impressive array of evidence to back up his position. I invite people to read the case that he provides as to why these statements are wrong, and make up their own minds. The point I make here is that Plimer is able to draw on a substantial literature which simply cannot be ignored if we are to treat the subject matter seriously. We should attend to alternative views in their strongest form before we jump onto populist bandwagons created by those who may well have a vested interest in what they promote.

---

330 To read the case Plimer makes I invite the interested reader to read Ian Plimer, *Heaven + Hell*, Ballarat, Connor Court Publishing, 2015, 97-115.

Chapter 6

In-Church Responses to *Laudato Si'*

Political activists within the Catholic Church in Australia have not been slow to harness *Laudato Si'* to their preferred causes. Of course, the way this encyclical has been written certainly gives political activists every opportunity to make the best of what must seem like a "God-given" opportunity. Nevertheless, encyclicals are not meant to be used in this crude and unsophisticated way. In education circles *Laudato Si'* should be used for informing students about Catholic social teaching and its binding nature. But educators need to be clear about what in this Encyclical is binding Catholic teaching and what is not.

Moreover as Saint John Paul II reminded us, Catholic social teaching "belongs to the field, not of ideology, but of theology and particularly of moral theology".[331] [Emphasis added]

So, too, charitable agencies need to be very careful that their very good works which accord with Catholic social teaching should not be confused with their partisan views about how the world should best be changed politically and economically. Moreover, selectively choosing which bits of an encyclical they wish to emphasise in promoting their laudable work for social justice in certain areas is not an excuse for bypassing their responsibility where the most fundamental social justice questions are concerned.

---

331 *Sollicitudo rei socialis*, n. 41.

Catholic Earthcare Australia

Catholic Earthcare Australia describes itself in these terms:

> Catholic Earthcare Australia is the ecological agency of the Australian Catholic Bishops' Conference. We were established in May 2002 and received our present Mandate in 2003.
>
> Our vision is for an ecologically sustainable and resilient Australia, where Catholic communities play an active part in the holistic care of social, human and environmental ecology.[332]

Catholic Earthcare says this about *Laudato Si'*: "*Laudato Si'* has the potential to unite and mobilise the global Catholic community."[333]

It is worth taking some time to examine Catholic Earthcare's response to the Encyclical, a response which uncritically advocates all the positions adopted in the Pope's Encyclical and, in some cases, goes beyond those positions.

Indeed there is something of a "leftist ultramontanism" (similar to "conservative ultramontanism") which treats every opinion expressed in this Encyclical that agrees with the "progressive" agenda as beyond criticism.

On its website this agency provides a number of resources[334] to guide people as to the way it thinks Catholics should receive the Encyclical and participate in the "forthright and honest debate" for which Pope Francis has called. Among its extensive provision of resources there is a very useful document titled "*Laudato Si' Summary*".[335]

That summary provides a clear account of the Encyclical, valuable for anyone wanting a general introduction to the material before attending to the whole document. A supporting document titled "Questions and Answers dealing explicitly with the Encyclical" attempts to answer some, but by no means all, of the questions raised by the Encyclical, answers which often themselves

---

332 http://catholicearthcare.org.au/about/
333 http://catholicearthcare.org.au/ecological-encyclical/
334 ibid.
335 http://catholicearthcare.org.au/wp-content/uploads/2015/05/Encyclical-Summary-EN.pdf

raise even more questions for discussion.

Notwithstanding the clear mismatch between the Pope's description of ecological issues faced by the contemporary world and what is set out in official UN Reports on the state of play, *Catholic Earthcare*, together with *Catholic Religious Australia*, and *Catholic Mission* reiterate the contentious factual claims made in *Laudato Si'*. These three Australian Catholic agencies published *The Francis Effect II –On Care for our Common Home*. At the beginning of this publication theologian Dennis Edwards states:

> *Laudato Si'* begins with a clear-eyed discussion of what is happening to our planet ... In particular Pope Francis offers a careful analysis of major issues we face, particularly pollution and global warming, the looming crisis of fresh water, and the loss of biodiversity, along with the decline in the quality of human life, breakdown of society, and global inequality.[336]

It will be noted that the very issues which Edwards claims to reflect a "careful analysis" of current environmental issues are those claims which I have shown to be among the most contentious in the Encyclical. But for a purely ideological use of the Encyclical, 22-year old Youth Engagement Officer for *Catholic Earthcare*, Terese Corkish, takes the game to a whole new level:

> Throughout the entire Encyclical I was reminded of Naomi Klein, Canadian author and activist. She is vehemently anti-capitalist, something I have become less and less afraid of recently. For me being anti-capitalist is about being realistic about human, economic and planetary limitations. It is about acknowledging the intrinsic values of places like the Great Barrier Reef, and endangered species.[337]

---

336 Denis Edwards, "Hope for our Common Home", *The Francis Effect II –On Care for our Common Home*, North Sydney, Catholic Mission, Catholic Religious Australia, and Catholic Earthcare Australia, 10.

337 Terese Corkish, "Young People's Call to Transform", *The Francis Effect II – On Care for our Common Home*, North Sydney, Catholic Mission, Catholic Religious Australia and Catholic Earthcare Australia, 48.

But is Corkish's admiration for Klein justified? "Klein is a pro-abortion, anti-corporate, anti-free enterprise agitator with no expertise in science or economics."[338] Indeed she has no completed university degree but she did study "philosophy and literature".[339] Her support for the pro-choice position on abortion is the exact antithesis of what Pope Francis regards as central to Catholic social teaching. She has, among other false claims, wrongly claimed that the "Medieval Warm Period was thoroughly debunked long ago."[340] She is no real authority on the subject, no matter that she was called in to advise Pope Francis. And nowhere in Catholic social teaching are we expected to canonise ideologies that semi-divinise nature. Judaeo-Christian thought demythologised nature. "While continuing to admire its grandeur and immensity, it no longer saw nature as divine."[341] Acknowledging the duty of human beings to care for nature, more importantly we have to "protect mankind from self-destruction."[342]

The temptation for Catholic political activists such as Corkish is to allow political ideology to gazump both fact and Catholic social teaching while claiming to speak on behalf of "young people". In the first place, no one gets to speak on behalf of "young people" in general (any more than anyone can speak for "women" or "workers" in general). Today's young people are intelligent persons who express just as much diversity of opinion among themselves as do other groups within society. In the second place, and in spite of Saint John Paul II's warning, we find ideology being confused with the Gospel the Church is called to proclaim.

> The strong link between climate change and justice is one that needs to be shouted from every pulpit and institution, the message – 'if you respect the dignity of life, you should be

---

338 Maureen Malarkey, "Naomi Klein", *First Things*, https://www.firstthings.com/tag/naomi-klein
339 http://www.c-span.org/video/transcript/?id=8205
340 Ian Plimer, *Heaven + Hell*, Ballarat, Connor Court Publishing, 2015, 42-44.
341 *Laudato Si'*, n. 78.
342 *Caritas in veritate*, 29 June 2009, 51 and quoted in *Laudato Si'*, n.79.

fighting climate change' is, and should be central to Catholic social teaching.[343]

It is untrue to state that "climate change is central to Catholic social teaching" and Pope Francis does not say that. In fact, he says that "the Church does not presume to settle scientific questions or to replace politics".[344] Moreover, it simply does not follow from the doctrine of the intrinsic dignity of the human person that one must "fight climate change". It is possible to be faithful to Catholic social teaching and justifiably consider either that climate change is a natural phenomenon or that a particular way of "fighting" it (eg controlling population growth by abortion, contraception and the like) is immoral. The Gospel of Jesus Christ is central to Catholic social teaching which itself embraces a far wider consideration of environmental issues than climate change. And if "fighting climate change" turns out to be no more than tilting at windmills, the young people open to the Church will have been sold a pup. It is not the Church's mission to solve scientific disputes, let alone recommend a raft of economic and other policies recommended by the activist few as not only necessary "to save the planet" but also, apparently, "necessary for salvation" such that it must be preached from every pulpit.

Corkish has overlooked the fact that Pope Francis, despite his clear and, I would say, exclusive preference for one particular scientific view, nevertheless still acknowledges that:

> there are certain environmental issues where it is not easy to achieve a broad consensus. Here I would state once more that ***the Church does not presume to settle scientific questions or to replace politics***. But I am concerned to encourage an honest and open debate so that particular interests or ideologies will not

---

343 Terese Corkish, "Young People's Call to Transform", *The Francis Effect II – On Care for our Common Home*, North Sydney, Catholic Mission, Catholic Religious Australia and Catholic Earthcare Australia, 50.

344 *Laudato Si'*, n. 188.

prejudice the common good.³⁴⁵ [Emphasis added]

To be sure the Holy Father thinks there is a "broad consensus" on climate change science, but even that does not mean the Church has the competence to resolve the scientific controversies around this form of climate science, let alone preach it as a core doctrine of the Gospel.

Corkish then goes on to make this claim based upon no provided evidence:

> We see people who will protest against abortion clinics but who will not turn up to a climate march or even recycle, and Pope Francis is making significant steps to make the link between green values and basic human dignity.³⁴⁶

This comment is completely unsubstantiated. Possibly there are Catholics who see their apostolate primarily in protesting against abortion, just as there are Catholics who see their apostolate to be attending climate marches or recycling. What ideological point is Corkish attempting to score? Given that green values embrace population control, abortion³⁴⁷, and the like, how can she possibly assert that Pope Francis wants to make a link between "green values" and basic human dignity?

This kind of unpleasant posturing generates more heat than light, and not the kind of debate Pope Francis calls for.

> Since everything is interrelated, concern for the protection of nature is also incompatible with the justification of abortion. How can we genuinely teach the importance of concern for other vulnerable beings, however troublesome or inconvenient they may be, if we fail to protect a human embryo, even when its presence is uncomfortable and creates difficulties? "If personal and social sensitivity towards the acceptance of the new life is

---

345 *Laudato Si'*, n. 188.
346 *Ibid.*
347 The political manifesto of the Australian political party *The Greens,* supports "the right to choose" abortion as stated a*t* http://greens.org.au/sites/greens.org.au/files/election_platform_screen.pdf

lost, then other forms of acceptance that are valuable for society also wither away".[348]

But, much more to the point is that opposing abortion is central to Catholic social teaching while there is no such infallible teaching requiring Catholics to oppose climate change in the manner she prescribes. As Saint John Paul II pointed out in *Evangelium vitae*:

> What is urgently called for is a general mobilization of consciences and a united ethical effort to activate a great campaign in support of life. *All together, we must build a new culture of life: new, because it will be able to confront and solve today's unprecedented problems affecting human life; new, because it will be adopted with deeper and more dynamic conviction by all Christians; new, because it will be capable of bringing about a serious and courageous cultural dialogue among all parties.* While the urgent need for such a cultural transformation is linked to the present historical situation, it is also rooted in the Church's mission of evangelization. The purpose of the Gospel, in fact, is "to transform humanity from within and to make it new". *Like the yeast which leavens the whole measure of dough (cf. Mt 13:33), the Gospel is meant to permeate all cultures and give them life from within, 124 so that they may express the full truth about the human person and about human life.*[349]
> [Emphasis added]

Although Pope Francis prefers the so-called "consensus position" on global warming, he also promotes the teaching of Saint John Paul II when it comes to prolife issues and the centrality of the family in Catholic social teaching. As I pointed out earlier in this chapter Pope Francis reasserted traditional Catholic teaching by putting the prolife issue as a central feature of his *Laudato Si'*.

> In the face of the so-called culture of death, the family is the heart of the culture of life". In the family we first learn how

---

348 *Laudato Si'*, n. 120. The last sentence of this paragraph is referenced to Encyclical Letter *Caritas in Veritate* 29 June 2009, 28.
349 *Evangelium vitae*, n. 95.

to show love and respect for life; we are taught the proper use of things, order and cleanliness, respect for the local ecosystem and care for all creatures. In the family we receive an integral education, which enables us to grow harmoniously in personal maturity. In the family we learn to ask without demanding, to say "thank you" as an expression of genuine gratitude for what we have been given, to control our aggressivity (*sic*) and greed, and to ask forgiveness when we have caused harm. These simple gestures of heartfelt courtesy help to create a culture of shared life and respect for our surroundings.[350]

In fact, where Corkish berates prolife Catholics for failing to embrace the green ideology, Pope Francis actually criticises environmentalists for failing to embrace the pro-life message at the heart of Catholic Social Teaching. This **"preferential option for human life"**[351] is further expressed in *Laudato Si'* thus:

> On the other hand, it is troubling that, when some ecological movements defend the integrity of the environment, rightly demanding that certain limits be imposed on scientific research, they sometimes fail to apply those same principles to human life. *There is a tendency to justify transgressing all boundaries when experimentation is carried out on living human embryos. We forget that the inalienable worth of a human being transcends his or her degree of development. In the same way, when technology disregards the great ethical principles, it ends up considering any practice whatsoever as licit.* As we have seen in this chapter, a technology severed from ethics will not easily be able to limit its own power.[352] [Emphasis added]

Proponents of "green politics" in the Catholic bureaucracy need to be fair to what the Pope has actually said in relation to infallible Catholic social teaching and not suggest that Papal opinions on global warming and its causes have a level of authority which the Pope himself does not claim. The condemnation of abortion is

---

350 *Laudato Si'*, n. 213 citing John Paul II, *Centesimus Annus* (1991), 39.
351 My expression, see above.
352 *Laudato Si'*, n. 136.

settled Catholic social teaching. Opinions on global warming are subject to revision in the light of new information provided by the relevant sciences involved.

> Given such unanimity in the doctrinal and disciplinary tradition of the Church, Paul VI was able to declare that this tradition [re abortion] is unchanged and unchangeable. Therefore, by the authority which Christ conferred upon Peter and his Successors, in communion with the Bishops-who on various occasions have condemned abortion and who in the aforementioned consultation, albeit dispersed throughout the world, have shown unanimous agreement concerning this doctrine-I declare that direct abortion, that is, abortion willed as an end or as a means, always constitutes a grave moral disorder, since it is the deliberate killing of an innocent human being. This doctrine is based upon the natural law and upon the written Word of God, is transmitted by the Church's Tradition and taught by the ordinary and universal *Magisterium*.[353] [Emphasis added]

This infallible declaration of the "grave moral disorder" of the crime of abortion is nowhere matched to any degree whatsoever with regards to 'green politics" in general or "climate change" in particular.

Political Misuse of the Encyclical

The failure of many in the Catholic education system to distinguish between matters fundamental to faith and other more tangential matters has been well exemplified in their response to *Laudato Si'*. Indeed, the Encyclical has been put in service of promoting the "progressive" and "green" agendas of some schools and their teachers.

In a media release dated the 24th of November 2014, "Canberra students urge PM to have a heart over climate", the use of students as political campaigners and demonstrators is celebrated by *Caritas*

---

[353] *Evangelium vitae*, n. 62. Saint John Paul II also referenced The Second Vatican Council's Dogmatic Constitution on the Church, *Gaudium et spes*, n. 25 at the end of this paragraph.

Australia. The "chief" of *Caritas*, Paul O'Callaghan, makes this statement:

> Young Australians want to see real action on climate change. Together they are sending a clear message [to] our Prime Minister ahead of the Paris climate talks later this month.
> Climate change is the single biggest threat to reducing global poverty.

This is climate politics, pure and simple. O'Callaghan does not "know" what "young Australians" think about climate change. What is really happening here is that his organisation is running a campaign and has been encouraging students to participate in that campaign, students most of whom are not old enough to vote.

The political programme is named the *Our Common Home* campaign, and the part involving school children "Heart4Climate". Then we are told 60 students from Daramalan College in Canberra "joined Caritas Australia and thousands of students from Catholic Schools across Australia in calling for strong climate change action ahead of climate talks in Paris."

A number of issues attend this *Caritas* campaign.

1. *Laudato Si'* has been used as a political manifesto;
2. School children in Catholic Schools have been drafted into a letter writing political campaign;
3. 60 Daramalan College students were used for a "photo opportunity in front of Parliament House"; and
4. These students then presented "thousands of heart shaped letters to the Prime Minister's Office."[354]

So when Pope Francis says that there is a "need for forthright and honest debate", what *Caritas* and some Catholic schools give us is a political campaign in which those involved have already made up their minds and see no "need for forthright and honest debate". But schools are meant to be places where students are enabled to pursue the truth in an as objective way as possible considering all relevant

---

354 CathNews, "Canberra students urge PM to have a heart over climate", 24 November 2015.

facts and the arguments used based upon those facts. Students need to be given the space for a genuine debate on the various opinions about the causes of climate change and what should be the morally correct human response. It is reasonable to ask whether, and to what extent, and in what manner, have alternative views been presented to students. In chapter Five I mentioned scholars like Ian Plimer, Robert M Carter, Garth Paltridge, and Roy Spencer. Each of these books directs attention to the vast literature on the issue of climate change. The danger of students being urged into political action is that they will be unduly influenced by educators who privilege one school of thought over another even to the point of not even mentioning them at all. In which case it is more a case of students having their minds already made up for them.

The political campaign in which *Caritas* and students are involved rests on the proposition that "Climate change is the single biggest threat to reducing global poverty". But this is just an unsubstantiated claim and by no means an obviously compelling truth claim in the context of a reduction in global poverty at a time of global warming.

Caritas Australia

Just what is the political, social and moral agenda of *Caritas Australia*?

> *Caritas Australia* sits within the auspices of the Australian Catholic Bishops Conference (ACBC) and operates in accordance with ACBC policy and mandate.[355]
>
> *Caritas* is calling for the world to put the poor first and ensure the safe future of the planet by uniting behind the United Nations' new Sustainable Development Goals which promise to end extreme poverty, tackle inequality and take action on climate change by 2030.[356]

---

355 http://www.caritas.org.au/about/at-a-glance
356 http://www.caritas.org.au/learn/sdg

So, *Caritas Australia* is at one and the same time both a Catholic agency and an agency committed to the United Nations' Sustainable Development Goals, some of which are clearly in conflict with Catholic social teaching as set out in *Laudato Si'*, specifically important parts of goals 3 and 5.[357]

Caritas Australia seems to be giving unqualified support to these goals. But that includes a commitment to ensure by 2030 that there is "universal access to sexual and reproductive health care services, including for family planning, information and education, and the integration of reproductive health into national strategies and programmes." That is to say, universal access to abortion, contraceptives and the like, all of which have been condemned by Pope Francis. "Since everything is interrelated, concern for the protection of nature is also incompatible with the justification of abortion."[358]

Why would *Caritas Australia* give unqualified support to the Sustainable Development goals which include support for abortion and contraception, when the Catholic Church, under whose banner it flies, has declared them to be immoral practices which are intrinsically evil? The Church teaches this:

> Since the first century the Church has affirmed the moral evil of every procured abortion. This teaching has not changed and remains unchangeable. Direct abortion, that is to say, abortion willed either as an end or a means, is gravely contrary to the moral law.[359]
>
> In the case of an intrinsically unjust law, such as a law permitting abortion or euthanasia, it is therefore never licit to obey it, or to *"take part in a propaganda campaign in favour of such a law, or vote for it"*.[360] [Emphasis added]

---

357 These goals were discussed earlier in this book in chapter 5.
358 *Laudato Si'*, n. 120.
359 *Catechism of the Catholic Church*, n. 2271.
360 *Evangelium vitae*, n. 72 and cf Congregation for the Doctrine of the Faith, *Declaration on Procured Abortion* 18 November 1974, n. 22.

On the other hand, there is no infallible teaching on climate change or climate change politics. And yet we find *Caritas Australia* propagating the Sustainable Development Goals with its commitment to abortion and campaigning for climate change action of a particular kind. *Laudato Si'* is not to be used as a political manifesto for political action. And the more so when one is campaigning for political action in favour of another manifesto (Sustainable Development Goals) which promotes abortion and contraception contrary to Catholic teaching and which is singled out by Pope Francis as incompatible with integral ecology in his Encyclical.

As I have demonstrated earlier, support for abortion and contraception is an essential part of the population control package. Does *Caritas Australia* support population control? It is not clear that the answer to that question is in the negative. But Pope Francis said this:

> Instead of resolving the problems of the poor and thinking of how the world can be different, some can only propose a reduction in the birth rate. At times, developing countries face forms of international pressure which make economic assistance contingent on certain policies of "reproductive health". Yet "while it is true that an unequal distribution of the population and of available resources creates obstacles to development and a sustainable use of the environment, it must nonetheless be recognized that demographic growth is fully compatible with an integral and shared development".[361]

CAFOD UK

The *Catholic Agency for Overseas Development* (CAFOD) is the official aid agency of the Catholic Church in England and Wales and part of *Caritas International*.[362] It has devoted a great deal of

---

[361] *Laudato Si'*, n. 50, and cf Pontifical Council for Justice and Peace, Libreria Editrice Vaticana, 2004, *Compendium of the Social Doctrine of the Church*, 483.
[362] http://www.cafod.org.uk/

space on its website to *Laudato Si'* because it is a Catholic agency like *Caritas Australia*. And like *Caritas Australia*, CAFOD also advocates for the Sustainable Development Goals in an unqualified and uncritical manner. Under the heading of Sustainable Development Goals, they say:

> The SDGs are universal, meaning they are equally applicable to all countries. They include challenging targets for rich countries as well as poor.[363]

No hint here that these same development goals contain a commitment to "reproductive health", "reproductive rights", and "gender equality" and therefore to promoting greater availability of contraception and "safe" abortion.

CAFOD then gives us the whole sustainable development programme by encouraging people to:

> Commit to implementation of all the global goals and include this in national development plans and priorities. Universality is one of the keys for successful implementation at both global and national level. It should address not only the achievement of the goals within that country but also each country's fair contribution to global achievement of the goals.[364]

CAFOD makes great play of its "Catholic" credentials and its commitment to Catholic social teaching. And a lot of what it stands for is good and, like *Caritas Australia*, what a lot of what it does on the ground is very good indeed. But this is seriously marred by the reality that CAFOD appears to be highly selective about what it chooses from Catholic social teaching in general and from *Laudato Si'* in particular. While Pope Francis sets his face against the population control programmes of the UN and other bodies, CAFOD appears to endorse it when it says it wants everyone to *"commit to the implementation of all the global goals"* contained in the Sustainable Development Goals.

---

[363] file:///C:/Users/John/Documents/Campion%20docs/BOOK%20RE%20LS%20(2)/CAFOD_SDG_Action_Towards_2030_%20Single.pdf
[364] *Ibid.*

CAFOD, as is its right, is committed to the "political consensus" position on climate change and the steps it believes should be taken to ameliorate the warming of the planet. Be that as it may, I repeat what I said earlier in relation to *Caritas Australia*: there is no infallible teaching on climate change or climate change politics,

Yes, Pope Francis believes in the "political consensus" where climate change is concerned. But he also teaches just why the population controllers are wrong.

> At times, developing countries face forms of international pressure which make economic assistance contingent on certain policies of "reproductive health". Yet "while it is true that an unequal distribution of the population and of available resources creates obstacles to development and a sustainable use of the environment, it must nonetheless be recognized that demographic growth is fully compatible with an integral and shared development".[365]

Pope Francis continues to reiterate and teach the wrongfulness of abortion.

> It is horrific even to think that there are children, victims of abortion, who will never see the light of day. Unfortunately, what is thrown away is not only food and dispensable objects, but often human beings themselves, who are discarded as unnecessary.[366]

St Vincent de Paul Society Australia

The St Vincent de Paul Society, affectionately known throughout Australia as "Vinnies", has an enviable record of care for the poor and those at risk. From my own personal experience I know the good that they do by "serving Christ in the poor with love, respect, justice, hope and joy, and by working to shape a more just and

---

365 *Laudato Si'*, n.50.
366 BBC News, "Pope Francis denounces 'horror' of abortion", 14 January 2014, http://www.bbc.com/news/world-europe-25723422

compassionate society".[367] Their "Key Values" are "commitment, compassion, respect, integrity, empathy, advocacy, and courage", all of which are elaborated on their website.[368]

And yet Vinnies promotes and privileges one particular point of view on anthropogenically caused global warming. The first thing that needs to be asked is how this issue is connected with its Mission and Vision. Vinnies claims no expertise in science or scientific disputes, let alone public policy decision making in this area. Yet Vinnies take on the issue in the Summer 2007-2008 edition of *The Record*, which is still prominently displayed on their website.[369]

> For many of us in respect of climate change, the turning point has been the realisation that the desire for justice for the poor cannot be fulfilled if global warming is allowed to continue unabated, for it is the poorest who are in the front line as water levels rise and crops are affected by temperature changes.

This is hardly a justification for Vinnies' involvement in this issue. The references to rising levels of water and crops being affected by temperature changes are opinions for which no justification is provided. Moreover, it assumes that human beings have caused the climate to change and have the power to turn around the drift to higher temperatures. Notwithstanding that no satisfactory explanation is provided as to why Vinnies ought to take a particular position with regard to climate change, it enters the battle with a particularly apocalyptic article on the subject by Fr Sean McDonagh SCC. McDonagh begins his article with the following assertions:

> In 2007 Australia was in the midst of a 1000-year drought which is, most probably, due to global warming. Will there be enough water to support the population of Perth or Sydney?
>
> A report from the International Panel on Climate Change (IPCC) in April 2007 predicts that "Australia will be hit by more

---

367 From the Mission Statement of St Vincent de Paul, https://www.vinnies.org.au/page/About/MissionVision/
368 *Ibid.*
369 https://www.vinnies.org.au/icms_docs/172581_The_Record_Summer_2007-2008.pdf

> frequent and intense heat waves, bushfires, floods, droughts and landslides as global warming causes the temperature to rise during this century."
>
> The IPCC report also predicts that a rise in sea levels is virtually certain to cause greater coastal inundation, erosion and loss of wetlands. A significant rise in sea levels will inundate many of the cities of the world and create a torrent of environmental refugees. A rise of one metre in the sea level would make it impossible for over 30 million Bangladeshis to live in the delta area.
>
> Climate change will cause horrendous pain to hundreds of millions of people, which is why it ought to be one of the priorities of an organisation like the St Vincent de Paul Society, whose members are dedicated to helping those who are less well off in our world.[370]

Each and every one of these assertions of "fact" is open to serious debate. Indeed since this article was first published a great deal of discussion has gone on touching on the matters raised.

The claim of a 1000 year drought is particularly unsubstantiated and unsubtantiatable. Over the last 150 years there have been 10 major droughts in Australia. Droughts are part of the natural environment in Australia. The last major drought was from 2002-2007.

> This period from 2002 to 2007 ranks with the Federation Drought and the Forties Drought as one of the three most severe, widespread and prolonged dry periods since 1900.[371]

According to this Australian Government website, "The experience of natural disaster has come to be seen as part of the Australian national character as described in the poem 'My Country' by Dorothea McKellar (1904)".

---

370 Sean McDonagh, "Ethics and Climate Change", *The Record*, Summer 2007-2008, St Vincent de Paul Society, https://www.vinnies.org.au/icms_docs/172581_The_Record_Summer_2007-2008.pdf

371 http://www.australia.gov.au/about-australia/australian-story/natural-disasters

> I love a sunburnt country, a land of sweeping plains,
> Of ragged mountain ranges, of droughts and flooding rains.
> I love her far horizons, I love her jewel-sea,
> Her beauty and her terror - the wide brown land for me!

As Ian Plimer has pointed out[372], instead of being guided by the advice from their own government department dealing with historical information, Australian governments allowed themselves to be panicked into spending billions of dollars building desalination plants in face of the doomsday predictions that the drought would continue indefinitely only to see them mothballed before they could be used because the rains came and the reservoirs were once again replenished.

The point here is that being stampeded into wasting vast sums of money because of doomsday predictions actually has the potential to disadvantage the poor by diverting money away from projects that might have been spent more wisely on measures to alleviate their poverty.

Where advancing sea levels are concerned, I again refer the interested reader to an alternative account of the science which may be found in Plimer's *Heaven and Hell*.[373]

McDonagh's apocalyptic prediction that "climate change will cause horrendous pain to hundreds of millions of people" is in conflict with the data reported by reputable UN bodies as outlined in chapter 5. It simply does not advance the "forthright and honest debate" for which Pope Francis calls if the emotional temperature of the debate is raised to this level. Nor will it assist a mature debate about the means needed to address the improving but still unacceptable level of poverty, particularly in developing countries.

It is difficult to understand why Vinnies should accept advice from such an extreme quarter even if it was within Vinnies' competence to make these sorts of judgements where the science

---

372 Ian Plimer, *Heaven and Hell*, Ballarat, Connor Court Publishing, 2015, 101-105.
373 *Ibid.*, 99-100.

of climate change is concerned. In any case, Vinnies are different to Caritas. They work to alleviate local poverty, as compared to global poverty. It is also hard to see a direct connection between helping the poor on our streets and any possible threats from climate change, humanly or naturally caused.

Conclusion

Father George Rutler and George Cardinal Pell are other critics of the parts of *Laudato Si'* already referred to in this paper. For Rutler:

> Pope Francis' encyclical on the ecology of the earth is adventurously laden with promise and peril. It can raise consciousness of humans as stewards of creation. However, there is a double danger in using it as an economic text or scientific thesis. One of the Pope's close advisors, the hortatory Cardinal Maradiaga of Honduras said with ill-tempered diction: "The ideology surrounding environmental issues is too tied to a capitalism that doesn't want to stop ruining the environment because they don't want to give up their profits." From the empirical side, to prevent the disdain of more informed scientists generations from now, papal teaching must be safeguarded from attempts to exploit it as an endorsement of one hypothesis over another concerning anthropogenic causes of climate change. It is not incumbent upon a Catholic to believe, like Rex Mottram in Brideshead Revisited, that a Pope can perfectly predict the weather.[374]

In another place Rutler makes these further observations. He refers to the astronomer (and quite likely a Catholic priest as well) Nicolaus Copernicus (1473-1543), who dedicated his prime text *On the Revolution of the Celestial Orbs* (*De revolutionibus orbium coelestium*) to Pope Paul III.

> Here the Church nursed science when the Protestant leaders were condemning anything that did not accord with their reading of Scripture. Martin Luther had called Copernicus "that madman

---

[374] Father George W Rutler, "Mixing Up the Sciences of Heaven and Earth" *Crisis Magazine*, 18 June 2015.

[who] wants to throw the art of astronomy into confusion" by denying that Joshua told the sun, and not the earth, to stand still. The Spanish theologians Diego De Zúñiga and Melchior Cano invoked against Protestant literalists the Augustinian exegetical principle that excluded the human language of Scripture from scientific proof texts. Though a pious Lutheran, the heliocentrist Johannes Kepler, shunned by his co-religionists, found friends among the Jesuits and had the honour of being plagiarized by the Catholic Galileo. *Pope Urban VIII, somewhat offended when he sensed that his protégé Galileo had satirized him as "Simplicio" in his "Dialogue Concerning the Two Chief World Systems", patiently urged Galileo to stay on the right track of speculation, and not to declare theory a fact.* [Emphasis added

It may reasonably be said that Galileo was right and wrong, and so were some of his opponents, among whom St. Robert Bellarmine did not distinguish himself. Right in asserting the motion of the earth, which opponents denied, Galileo was wrong about the "solar stasis", or immobility of the sun, which his opponents accepted. *Both succumbed to error when they paraded theory as fact and scorned opponents as "deniers."* That alchemy of pride turns science into a false cult of scientism, which is unscientific science, while clerics abusing their authority descend to a false cult of clericalism, which is irreligious religion. A salutary example of how to order things rightly by humility was Christopher Clavius, the German Jesuit astronomer and mathematician, one of the commissioners for the Gregorian calendar, revered throughout Europe, who was a firm geocentrist. Telescopic observations with Galileo changed his mind, albeit with reservations, and he remained aloof from polemics.

... In the saga of environmentalism, the eleventh century Anglo-Scandinavian King Canute is often mistakenly evoked as a symbol of arrogance for setting his throne up on an English beach, possibly at Westminster or West Sussex or Southampton, and ordering the tides to roll back. The details are vague, but the real point of the story is that Canute deliberately choreographed that drama to instruct his flattering courtiers in the limits of earthly power against the seas and skies. They had preened that their king could slow the rise of the oceans and heal the planet. The tides did not withdraw, the king and his court got wet and Canute declared: "Let all men know how empty and worthless is the power of kings, for there is none worthy of the name, but He whom heaven, earth, and sea obey by eternal laws." It was a warning for scientists flattered by clerics, and clerics flattered by scientists. King Canute's performance was better than any flamboyant light show. Better still, King Canute then placed his crown on the great crucifix in Winchester Cathedral and never wore it again. In matters of speculative science, it would be edifying to see the members of the Intergovernmental Panel on Climate Change, the directors of the World Bank Group, corporate executives, and academics, do the same.[375]

And Cardinal George Pell says this about *Laudato Si'*:

It's got many, many interesting elements. There are parts of it which are beautiful but the church has no particular expertise in science … the church has got no mandate from the Lord to pronounce on scientific matters. We believe in the autonomy of science," (Financial Times on Thursday July 16, 2015) … "the encyclical had been 'very well received'" …The Pope had "beautifully set out our obligations to future generations and our obligations to the environment."[376]

---

375 Fr George W Rutler, "Making Dogma Out of Unsettled Science", *Crisis Magazine*, 14th December 2015.
376 http://cathnews.co.nz/2015/07/24/cardinal-pell-questions-science-behind-laudato-si/

Chapter 7

And Finally

This Encyclical is a very long, even a "sprawling" document as George Weigel puts it. And "sprawling" seems to be the right adjective here.

The Pope attempts to take on an extremely wide canvas of human problems. But an encyclical is a medium incapable of adequately dealing with all the issues which this Encyclical attempts to cover. Even a very large book might not be adequate for the task. As a consequence, the major threats to human life identified earlier (contraception, sterilisation, and abortion), and their links to foreign aid, do not get the intensity of coverage that perhaps they should in this work. In fact contraception and sterilisation are not explicitly mentioned at all even though promoting such things is at the forefront of the policies of the population controllers.

However, there are other elements of the Encyclical which are superb and give sound moral advice to the world "at a moment when incoherence, skepticism, and nihilism dominate Western high culture, and when fanaticisms claiming various divine or quasi-divine warrants wreak havoc from northern Nigeria to the Levant to the Donbas".[377]

Nevertheless, one is right to be worried when popes take sides on scientific disputes, or alternatively, are unduly influenced by one side of a dispute, particularly when there is no acknowledgement of the sources relied on to privilege the side adopted. In the case of *Laudato Si'* we might ask: "Who is really speaking? – the Pope

---

377 Cf http://www.nationalreview.com/article/419933/parsing-popes-ecology-encyclical-its-about-lot-more-climate-change-george-weigel

or his expert advisers? Moreover, the Encyclical is also open to criticism about the accuracy of some of the claims it makes, and the inappropriate use of high rhetoric and extreme language, language guaranteed more to obfuscate than to clarify.

But the parts of the Encyclical which clearly repeat Catholic social teaching command acceptance by all Catholics, and serious consideration by all who make public policy decisions for the common good. And this Encyclical, following the examples of Pope Emeritus Benedict and Saint John Paul II, strongly inserts into the public discussion the ethical insights of the Catholic faith where environmental issues are concerned, and especially the "preferential option for human life" in the context of an integral ecology. As Cardinal Pell has put it, the Pope has "beautifully set out our obligations to future generations and our obligations to the environment."

So, what is my own view on these matters? Broadly speaking, I too would promote all those morally responsible programmes which enhance our stewardship of the environment, improve our efficiency in the use of resources by fighting to reduce waste and excess, reduce environmental degradation, and contribute to the renewal of planet earth. Responsibility for these issues together with the need to protect human rights, eliminate poverty wherever we can, and enhance education and health services globally, lies primarily with governments at all levels from local to national, and institutions and individuals in the way we conduct our lives.

If the "political consensus" around climate change is proved to be significantly in error, then major decisions could be made on a false premise leading to economic and social harm on a large scale. On the other hand, I freely acknowledge that if the scientific and political "consensus" proves right, we may come to rue the missed opportunity to avoid widespread economic and social harm. This is precisely why the present moment is a time for clear heads, not for bullying or apocalyptic scare-mongering. But the world has history where this is concerned as the neo-Malthusian prophecies of Paul Ehrlich and the Club of Rome in the 1960s and 1970s bear eloquent witness. And the Church, too, has history as evidenced by the Galileo saga.

What of issues around climate change? Unlike other environmental issues, this is still an area of limited knowledge and disagreement especially with regard to the causes of global warming. In the end scientists have to be true to their vocation and seek the empirical truth of the matter for its own sake and for the sake of humanity. Governments have to make judgements about what is being put to them by a whole range of people, and take such action as they see fit to discharge their responsibilities to human beings and the total environment in which we live.

Jeff Mirus, in his recent discussion about what is really meant by target setting to reduce temperature by 1.5 degrees C or 2 degrees C, makes these quite reasonable observations among others:

> Since we do not know all the causes of warming, and cannot control even some of those we do know, how can anyone guarantee that a particular target can be reached?
>
> Anthropogenic global warming may be real or it may be illusory. It may not be naturally self-correcting over time or then again it may be (as theories involving both sun spots and oceanic production of carbon dioxide both strongly suggest). It may or may not be the case that mankind can expect climate conditions to remain within such a narrow range of variability that the same land will be able to be used in the same way during every period of its history. This has not, in fact, been the case in the past, and both migrations and crop changes have occurred as a result.
>
> Still, the earth has (demonstrably) had significant variances in average temperature over the centuries. The current problem is closely related to low-lying countries, particularly island nations, because of rising sea levels and rising tides. I really do not know whether we can stick our collective finger in the dike, but I do know that oceanic islands come and go for a variety of reasons. (And as for the continental United States, as the thermal instability of Yellowstone goes, so goes much of the rest of the

country).[378]

It is, in my view, not the role of the Catholic Church to decide political and scientific matters. Those who have the scientific credentials are the ones who, in the end, must conduct the debate in a robustly scholarly manner seeking the truth of the matter and not politicising issues for other motives, *giving due regard as much for what scientists do not know as to what they do know.*

But the Church has the responsibility to articulate the fundamental moral principles at stake where the making of public policy is concerned, and to call to account those governments and international bodies who facilitate and promote policies based upon the violation of fundamental human values.

It is NOT that the Church should stay out of public debate on public policy issues as some of the authoritarian *bien pensants* would wish. On the contrary, the Church has both a right and a duty to contribute to the public conversation on any issues in which she judges she has the competence to speak. She must proclaim the Gospel of Jesus Christ with all of its implications to the world. That is to say, it is a case of the Church publicly speaking with authority where she has the competence to so speak, and offering any such other prudential advice as she sees fit while recognising that it is prudential judgement being exercised. And where the environment is concerned, Deus et natura non faciunt frusta! (God and nature do not work together in vain.)

The degeneration of public debate into a brawl with protagonists of a certain view of climate change using highly pejorative labels with which to silence opponents is not helpful. Use of terms such as "deniers", with all of its association with "holocaust deniers" has been particularly insidious. The search for truth is enhanced when well-credentialed experts who disagree with the "common view" are given their proper place in the debate. But global media has globally warmed the climate change debate by effectively closing

---

378 Dr Jeffrey Mirus, "Will a climate change goal change anything?" http://www.catholicculture.org/commentary/the-city-gates.cfm?id=1201

down intelligent public discussion. The colonising of the debate on climate by one group of scientists (a self-selected group of "climate scientists") to the exclusion of other relevant scientific disciplines is concerning. Equally concerning is the tendency of many in the scientific community of :

a) confusing theory with fact,
b) not being publicly honest about the limits of their knowledge, and
c) presenting themselves as moral experts when in fact in the area of moral philosophy they are almost always uninformed and unformed.

Again, it is not up to me to decide the competing views of scientists where climate is concerned. I do not have the expertise. But neither do those Church agencies which speak with magisterial certainty on things about which considerable uncertainty remains, and with hesitancy and even compromise on the certainties of Catholic social teaching by which all Catholics are bound and which Pope Francis enunciates in *Laudato Si'*.

This book has been written in full submission to Catholic social teaching but at the same time as a contribution to the "forthright and honest debate" for which Pope Francis has so clearly and emphatically called.

www.ingramcontent.com/pod-product-compliance
Ingram Content Group UK Ltd.
Pitfield, Milton Keynes, MK11 3LW, UK
UKHW041414180426
11947UKWH00007B/133